MATRICES AND MATHEMATICAL PROGRAMMING

An Introduction for Economists

Linear algebra and constrained optimisation play an increasingly important part in economics, econometrics and management studies. This book is an introduction to these subjects for students and research workers in economics and related fields.

The first eight chapters explain the core of matrix algebra and linear programming. There is considerable emphasis on the relation between the theory and practical methods of calculation. Chapter 4, on projections, will be particularly useful to students of econometrics. The discussion of linear programming emphasises duality throughout and carefully explains its economic interpretation. The last three chapters are an introduction to the theory of non-linear and concave programming: Chapter 10, on concave functions, should be a very helpful reference for courses in microeconomics. The static input-output model is discussed in an appendix.

The book is written in an easy and informal style, but due attention is paid to what follows from what—proofs are provided for all the main theorems, except in the last chapter. There are many examples and exercises. The book builds up from elementary beginnings to advanced topics usually omitted or glossed over by textbooks: these include the pseudo-inverse matrix, marginal valuation in linear programming and a careful treatment of constraint qualifications. The major prerequisite is school algebra, including the summation symbol and mathematical induction; the later chapters assume a knowledge of the rules of partial differentiation.

Matrices and Mathemathematical Programming is suitable as a textbook for courses in quantitative methods as early as the second year of an honours degree in economics. It should also be an ideal source of instruction for graduate students with a limited mathematical background.

Nicholas Rau is Lecturer in Political Economy at University College, London. He holds degrees in mathematics and economics from the University of Oxford and is the author of *Trade Cycles: Theory and Evidence.*

By the same author

TRADE CYCLES: Theory and Evidence

MATRICES AND MATHEMATICAL PROGRAMMING

An Introduction for Economists

NICHOLAS RAU

University College London

First published 1981 by
THE MACMILLAN PRESS LTD
London and Basingstoke
Companies and representatives throughout the world

ISBN 0–333–27768 6

Typeset by
Santype International Ltd
Salisbury, Wilts
and printed in Hong Kong

Contents

Preface

Two mathematical ideas play a prominent rôle in modern economics: the solution of systems of simultaneous equations and the maximisation of functions subject to constraints. They find application in such traditional areas of economic inquiry as price theory, public finance and international trade; in econometric testing and prediction; and in the activities of planners in the public and private sectors of rich and poor countries.

The purpose of this book is to give an account of these mathematical ideas at a level suitable for advanced undergraduates and beginning graduate students in economics and management. I hope also that research workers in economics and related fields will find it useful.

The main prerequisites for reading this book are a knowledge of school algebra, including the summation symbol and mathematical induction, and awareness of the difference between necessary and sufficient conditions. Section 7.3 uses partial-derivative notation and the last three chapters require the reader to be familiar with differential calculus: intuitive ideas of limits and continuity, ordinary and partial differentiation and exponential and logarithmic functions. The reader is not expected to know anything about integral calculus or rigorous real analysis: one proof in Section 9.2 uses some analysis, but the awkward step is indicated and can be omitted.

This is a book about the mathematics which economists use, and sometimes misuse; it is not a book about economics. I do work through the static Leontief model in Appendix D, several examples refer to 'consumers' and 'firms' and an economic example motivates the discussion of least-squares estimation in Section 4.3: that is all the economics this book contains, though the Notes on Further Reading at the end point to a wide variety of applications. I do not think that interesting economics can be taught by instant example in a book on mathematics and have no space to develop substantial arguments on (say) public finance or the theory of planning. But throughout the book I have attempted to keep in mind the needs of the student of economics, and the understandable reluctance of such students to devote time and effort to literature written for those interested in mathematics for its own sake.

So this is a book for the economist or management scientist who wishes to use mathematics, not for the prospective mathematician. But I do feel that economists who use mathematics should do so with care and intelligence. Most social scientists who apply mathematics, including myself, use 'canned formulae' at some stage in their work; but linear algebra and optimisation are so central to so much of modern economics that an introductory book should try to explain what follows from what. That is one reason (the other being the use of this book as a work of reference) why I have provided proofs of all the main theorems, except in the final chapter and in Appendix C. Some of the arguments in Chapter 9 are a bit rough, but generally I have tried to maintain a decent standard of rigour.

The mere provision of proofs is not enough to involve the reader in the logic of the argument. I have taken some pains to explain where we are heading and how the subject hangs together. The large number of examples are a mixed bag of illustrations of definitions, counterexamples and worked examples in the strict sense: all try to give the reader some notion of the limits of his knowledge. There are exercises at the end of most sections: the student should attempt them all.

Chapters 1–4 contain the main results on linear algebra which the economist needs, with two exceptions: determinants and eigenvalue theory. A brief survey of determinants is given in Appendix C. I have not covered eigenvalue theory because a thorough treatment of the subject requires complex numbers and makes little sense unless accompanied by applications to difference or differential equations; inclusion of these topics would have made the book far longer. The properties of projections and positive definite matrices which are used in elementary econometrics can be fully understood without eigenvalues. These are explained in Chapter 4.

Chapters 5–8 are on linear programming. The standard results are treated in detail, including the marginal interpretation of the dual, which has received less attention in textbooks than it deserves. My treatment of the simplex method is non-traditional: I motivate it in terms of duality and complementary pivoting rather than as a search over extreme points. I have found in my teaching that students have difficulties with a rigorous extreme-point approach, while a non-rigorous one leaves them confused about the connection between the algebra and the geometry.

Chapters 9–11 merely scratch the surface of nonlinear programming. All I have attempted is to get the first-order conditions right, starting with necessary conditions and then going on to the concave case. Many students of economics seem bad at recognising concave functions when they see them: this accounts for all the elementary calculus in Chapter 10. The examples in that chapter do not use Hessians: indeed, the whole book can be read without reference to Appendix C.

Appendices A and B fill in missing steps in the arguments of Chapters 6 and 7 respectively. Appendix C is on determinants and Appendix D on input–output, including the non-substitution theorem.

With regard to dependence of chapters, I should make it clear that my treatment of linear programming does not use much linear algebra. It would be possible to read Chapter 1, learn the pivoting procedure of Section 3.3, and then read Chapters 5–11, omitting only the part of Section 7.2 after Example 2. I have sometimes taken this route in teaching, and it is of course the recommended one for readers who know some linear algebra; but I think that the unaided reader who knows no linear algebra should try to understand linear equations before handling linear inequalities.

Section 3.3 is not needed until Chapter 6 but I think it should be read before Chapter 5: the development of linear programming should not be interrupted by a digression on an arithmetical trick. Chapter 4 is not used later in the text.

Appendix A can be read at any time after Section 6.2, Appendix B after Section 7.3. Most of Appendix C can be read at any time after Chapter 2, but the passage on symmetric matrices makes some reference to Chapter 4 and the final concavity test requires Chapter 10. The final part of Appendix D, on the nonsubstitution theorem, uses linear programming, in particular Section 8.2; the remainder requires only the first two sections of Chapter 2 and the vector inequality notation of Section 5.3.

A glance at the University College London calendar will explain why this book has eleven chapters. I wrote it after some years of teaching courses on Mathematics for Economics to third-year undergraduates, and on Quantitative Methods for beginning M.Sc. students, in the Department of Political Economy at UCL. Some of the graduate students were civil servants on secondment, with very little mathematical training; I hope that my conversations with them about their difficulties have helped to produce a simple and orderly exposition.

London, 1980 *Nicholas Rau*

Acknowledgements

This book has been rewritten more times than I am prepared to reveal: I am most grateful to Sheila Ogden for her splendid typing of numerous drafts. Dr. Malcolm Pemberton read the penultimate version with great care and made many useful suggestions and corrections. I should also like to thank Professor A. B. Atkinson, Professor John Black, Mr. W. J. Corlett and Dr. Joanna Gomulka for their helpful comments on parts of the manuscript. Any remaining errors are my responsibility.

Nicholas Rau

Notation

This book uses two old-fashioned conventions which I find useful. First, I refer to the reader in the male gender. Second, I use 'positive' to mean greater than zero, 'negative' to mean less than zero; 'non-positive' and 'non-negative' mean what they say; the terms 'strictly positive' and 'strictly negative' are not used. I apologise to readers who find the first convention offensive or the second unfamiliar.

Theorem 4 of Section 2.3 is stated and referred to as 'Theorem 4' within that section and referred to as 'Theorem 2.3.4' in other sections. A similar reference system is used for examples and exercises. Numbering of equations starts afresh each section: these are not referred to in other sections.

Vectors and matrices are printed in boldface type; the italic x_3 denotes the third component of the vector $\mathbf{x}$, whereas the boldface $\mathbf{x}_3$ denotes the third of a list of vectors $\mathbf{x}_1, \mathbf{x}_2, \ldots$. Square brackets are used when a component is selected from an implicit vector: thus $[\mathbf{Ax} + \mathbf{b}]_4$ denotes the fourth component of $\mathbf{Ax} + \mathbf{b}$.

Other notation will be explained in the text, but one innovation requires a few words here. This is the use of // to denote complementarity. Vectors $\mathbf{x}$ and $\mathbf{y}$ are said to be complementary if $x_i y_i = 0$ for each component i, in which case I write $\mathbf{x}//\mathbf{y}$. Readers who are used to the notation of the physical sciences may need reminding that this is not equivalent to orthogonality.

1 Vectors and Matrices

Most of this book is concerned with operations involving *lists* and *arrays* of numbers. This chapter introduces a notation called *matrix algebra* for expressing such operations.

After some comments on arithmetic and set notation in Section 1.1, we start our study of arrays in Section 1.2. The reader should make himself thoroughly familiar with the terms and manipulations of Sections 1.2 and 1.3 before moving on.

Section 1.4 is concerned with lists rather than arrays. The material of that section up to and including the proof of Theorem 3 is essential reading. The subsequent discussion of the Euclidean norm and vector geometry is not nearly as important, but we believe that many readers will find it helpful. Theorem 4 and the triangle inequality will be used only in Chapter 4 and in the last three chapters.

1.1 Numbers and sets

The *natural numbers* are 1, 2, 3 and so on. The *integers* are the natural numbers, their negatives (-1, -2, -3, ...) and zero. Thus a positive integer is the same as a natural number. *Rational numbers* are numbers of the form a/b where a is an integer and b a natural number. Clearly integers are rational numbers (let $b = 1$). The *real numbers* include the rational numbers, strange things like $\sqrt{2}$, even stranger things like π but nothing quite so strange as $\sqrt{-1}$ or ∞.

Having said all this, we must emphasise that the use of the word 'number' in this book is imprecise in two respects. The first concerns foundations. The remarks about 'strange things' were a deliberate evasion of any attempt to *define* a real number. Such a definition would be quite out of place in a book of this nature.

The second concerns jargon. This book draws on two branches of mathematics called *linear algebra* and *real analysis*. Chapters 2–4 are about linear algebra; nonlinear programming (Chapters 9–11) is an application of real analysis; and linear programming (Chapters 5–8) is somewhere in between. Now, it is common in linear algebra to refer to real numbers as *scalars* and to use 'number' to mean 'non-negative integer'. In real analysis it is common to use 'number' to mean 'real number' and to refer to integers as such. There are good reasons for these conventions but these will not concern us in this book. We shall typically use the algebraist's convention in Chapters 1–4 and the analyst's in other chapters, but we shall not be altogether consistent. In short, 'number' will mean 'real number' at some points and 'integer' at others, and it should be clear from the context which is meant.

It is important to note that infinity (∞) is not a real number. When the symbols ∞ and $-\infty$ are attached to the real number system we have the *extended real number system*, which will be used in the last section of this book and in Appendix B. Arithmetic in the extended real number system follows intuitively obvious rules. Thus

$$\infty > x > -\infty \text{ for any real } x$$
$$2 + \infty = 5 - (-\infty) = \infty$$
$$(-2) \times \infty = 5 \times (-\infty) = -\infty$$

and so on. The only difficulty lies with the expressions

$$\infty - \infty, \ \infty/\infty, \ 0 \times \infty.$$

The first two are undefined: they connote illegal operations, akin to dividing by zero. It is useful in many contexts to define $0 \times \infty$ to be zero, but we shall have no need of that in this book.

A popular mathematical shibboleth is: what is the square root of x^2? The answer, for any real number x, is that

$$\sqrt{x^2} = \begin{cases} x & \text{if } x \geq 0, \\ -x & \text{if } x < 0. \end{cases}$$

It is conventional to denote the non-negative number $\sqrt{x^2}$ by $|x|$ and call it the *absolute value* (or *modulus*) of x. Notice that $|x|$ is the greater of x and $-x$: this result may be written

$$|x| = \max(x, -x). \tag{1}$$

Two important properties of absolute values are

$$|x+y| \leq |x| + |y| \tag{2}$$

$$|xy| = |x| \cdot |y|. \tag{3}$$

We now prove (2) and (3). Observe that $x \leq |x|$ and $y \leq |y|$, so $x + y \leq |x| + |y|$. On the other hand, $-x \leq |x|$ and similarly for y, so $-x - y \leq |x| + |y|$. Applying (1) with x replaced by $x + y$, we obtain (2). Also $(xy)^2 = x^2y^2$: taking square roots, we obtain (3).

As we noted in the introduction to this chapter, the objects that will interest us in this book will not always be numbers. If it therefore helpful to have a language for dealing with collections of arbitrary objects.

A *set* is any collection of objects. Readers familiar with the first volume of Bertrand Russell's autobiography will know that this rather loose definition of a set leads to paradoxes in certain instances, but it is adequate for our purposes. Thus we may speak of the set of all economists in Europe, the set of all Asian currencies, and so on.

Let A be a set and let x be an object which belongs to A: we then write $x \in A$ (read: x *belongs* to A, or x is a *member* of A, or x is an *element* of A, or x is *in* A, or A *owns* x).

If two sets A and B are such that every member of A is also a member of B, we write $A \subset B$ (read: A is a *subset* of B, or A is *contained* in B, or B *contains* A). The sets A and B are said to be equal ($A = B$) if $A \subset B$ and $B \subset A$. If the set A is a subset of B but A is not equal to B, we say that A is a *proper subset* of B: thus A is a proper subset of B if and only if every member of A is a member of B and some member of B is not a member of A.

Example 1 Let P denote the set of British Prime Ministers in the years 1960–79. Then

$$\text{Wilson} \in P.$$

Let C denote the set of all Conservative Prime Ministers in those years and F the set of all female ones. Then F is a proper subset of C, and C is a proper subset of P.

If it is not too tiresome to enumerate all the elements of a set explicitly, we can describe the set by placing its elements within braces. Thus the set C of Example 1 can be written

{Macmillan, Home, Heath, Thatcher}

or as

{Heath, Home, Macmillan, Thatcher}.

Order is irrelevant in this context. The set F is

{Thatcher}

and is an example of a *singleton* (a set with only one member). Notice that the set $\{x\}$ is conceptually different from its member x.

A set is said to be *empty* if it has no members. For example, the set of all British Prime Ministers between 1960 and 1979 who belonged to the Liberal Party while in office is an empty set.

Observe that if A and B are sets such that A is empty, then $A \subset B$: for if A were not a subset of B, then A would own some object which is not in B and would therefore not be empty. It follows that if A and B are *both* empty, then $A = B$. Thus there is only one empty set: *the* empty set is denoted by $\varnothing$. Notice that the empty set is different from the number zero (keg beer is better than nothing; nothing is better than Paradise; 'therefore' keg beer is better than Paradise).

We can define sets by relations or properties. Let $\pi(x)$ be some statement involving a 'variable' x, not necessarily a number. Let A be a set. Then the subset of A consisting of all objects x in A for which $\pi(x)$ is true is written

$$\{x \in A : \pi(x)\}.$$

Example 2 Let F and C be as in Example 1. Then

$$F = \{x \in C : x \text{ is female}\}.$$

If J is the set of all integers then

$$\{x \in J : x^3 = x\} = \{-1, 0, 1\}$$
$$\{x \in J : 2x^2 - 5x + 2 = 0\} = \{2\}.$$

If R is the set of all real numbers then

$$\{x \in R : 2x^2 - 5x + 2 = 0\} = \{2, \tfrac{1}{2}\}.$$

We say that a set is *finite* if it has only a finite number of members, *infinite* otherwise. Thus the sets of Example 1 are finite, while the set of all integers is infinite.

Given two sets A and B we define the *intersection* of A and B to be the set of objects which belong to A and to B. This set is written $A \cap B$ (read: *A cap B*). We define the *union* of A and B to be the set of objects which are in either A or B or both. This set is written $A \cup B$ (read: *A cup B*). For example if $A = \{e, f, g, h\}$ and $B = \{g, h, i, j\}$ then $A \cap B = \{g, h\}$ and $A \cup B = \{e, f, g, h, i, j\}$.

It is easy to see that for any three sets A, B, C we have

$$(A \cap B) \cap C = A \cap (B \cap C)$$
$$(A \cup B) \cup C = A \cup (B \cup C).$$

These identities are known as the *associative laws* of intersection and union: their effect is to make expressions like

$$A \cap B \cap C \quad \text{and} \quad A \cup B \cup C \cup D \cup E$$

unambiguous. We shall come across other associative laws before very long.

Exercises

1. Prove that

$$||x| - |y|| \leq |x + y|.$$

2. Given the sets

$$X = \{1, 4, 7\}, \quad Y = \{2, 7, 9\}, \quad Z = \{1, 2, 5, 9\},$$

find $X \cap Y$, $Y \cap Z$, $X \cap Y \cap Z$ and $X \cup Y \cup Z$.

3. Notice that the expression $A \cup B \cap C$ is ambiguous and should therefore not be used. Give an example of three sets A, B and C for which

$$(A \cup B) \cap C \neq A \cup (B \cap C).$$

1.2 Matrix arithmetic

In Chapter 2 we shall be concerned with systems of linear equations of the form

$$\left.\begin{aligned} a_{11}x_1 + a_{12}x_2 + \cdots + a_{1n}x_n &= b_1 \\ a_{21}x_1 + a_{22}x_2 + \cdots + a_{2n}x_n &= b_2 \\ &\vdots \\ a_{m1}x_1 + a_{m2}x_2 + \cdots + a_{mn}x_n &= b_m \end{aligned}\right\} \tag{1}$$

Here $a_{11}, \ldots, a_{mn}, b_1, \ldots, b_m$ are $mn + m$ given scalars and $x_1, \ldots, x_n$ are the 'unknowns' for which we attempt to solve the system.

To deal with such systems, we shall find it convenient to develop a streamlined notation using matrices.

Let m and n be positive integers. An $m \times n$ (read 'm by n') *matrix* $\mathbf{A}$ is a rectangular array of numbers of the form

$$\mathbf{A} = \begin{pmatrix} a_{11} & a_{12} & \cdots & a_{1n} \\ a_{21} & a_{22} & \cdots & a_{2n} \\ & & \vdots & \\ a_{m1} & a_{m2} & \cdots & a_{mn} \end{pmatrix} \tag{2}$$

A $1 \times n$ matrix is called a *row vector* of *length* n. An $m \times 1$ matrix is called a *column vector* of *length* m. An $m \times n$ matrix is said to be *square* if $m = n$.

In (2), the row vector $(a_{i1} \cdots a_{in})$ is called the ith *row* of the matrix $\mathbf{A}$, for $i = 1, \ldots, m$. For $j = 1, \ldots, n$, the column vector

$$\begin{pmatrix} a_{1j} \\ \vdots \\ a_{mj} \end{pmatrix}$$

is called the jth *column* of $\mathbf{A}$. The number a_{ij} in the ith row and jth column of $\mathbf{A}$ is called the (i, j) entry of $\mathbf{A}$.

Let $\mathbf{p} = (p_1 \cdots p_n)$ be a row vector of length n, and let

$$\mathbf{q} = \begin{pmatrix} q_1 \\ \vdots \\ q_m \end{pmatrix}$$

be a column vector of length m. For $i = 1, \ldots, m$, the number q_i is called the ith *component* of $\mathbf{q}$: similarly for $j = 1, \ldots, n$, the number p_j is called the jth component of $\mathbf{p}$. Thus the (i, j) entry of an $m \times n$ matrix $\mathbf{A}$ may be regarded either as the jth component of the ith row of $\mathbf{A}$, or as the ith component of the jth column of $\mathbf{A}$.

Example 1 Let $\mathbf{A}$ be the 2×4 matrix

$$\begin{pmatrix} 6 & -1 & 3 & 7 \\ 2 & \frac{1}{2} & -4 & \frac{2}{3} \end{pmatrix}.$$

Then the $(1, 2)$ entry of $\mathbf{A}$ is -1, the $(2, 1)$ entry of $\mathbf{A}$ is 2, the $(2, 4)$ entry of $\mathbf{A}$ is $\frac{2}{3}$ and so on. The first row of $\mathbf{A}$ is the row vector

$(6 \ -1 \ 3 \ 7)$ of length 4. The second column of $\mathbf{A}$ is the column vector

$$\begin{pmatrix} -1 \\ \frac{1}{2} \end{pmatrix}.$$

Matrices are *ordered* arrays in the sense that two $m \times n$ matrices are deemed to be equal if and only if they are equal *entry by entry*. Given

$$\mathbf{A} = \begin{pmatrix} a_{11} & \cdots & a_{1n} \\ & \vdots & \\ a_{m1} & \cdots & a_{mn} \end{pmatrix}, \quad \mathbf{B} = \begin{pmatrix} b_{11} & \cdots & b_{1n} \\ & \vdots & \\ b_{m1} & \cdots & b_{mn} \end{pmatrix}$$

we have $\mathbf{A} = \mathbf{B}$ if and only if $a_{ij} = b_{ij}$ for all $i = 1, \ldots, m$ and all $j = 1, \ldots, n$. If

$$\mathbf{A} = \begin{pmatrix} 1 & 2 \\ 3 & 4 \end{pmatrix} \quad \text{and} \quad \mathbf{B} = \begin{pmatrix} 4 & 2 \\ 3 & 1 \end{pmatrix}$$

then $\mathbf{A}$ is *not* equal to $\mathbf{B}$.

We will now discuss operations on matrices.

Transposition

Let $\mathbf{A}$ be the $m \times n$ matrix

$$\begin{pmatrix} a_{11} & \cdots & a_{1n} \\ & \vdots & \\ a_{m1} & \cdots & a_{mn} \end{pmatrix}.$$

The *transpose* of $\mathbf{A}$, denoted by $\mathbf{A}^{\mathrm{T}}$, is the $n \times m$ matrix whose (j, i) entry is a_{ij} for $i = 1, \ldots, m$ and $j = 1, \ldots, n$. For example,

$$\text{if } \mathbf{A} = \begin{pmatrix} 1 & 2 & -1 \\ 4 & -5 & 0 \end{pmatrix} \text{ then } \mathbf{A}^{\mathrm{T}} = \begin{pmatrix} 1 & 4 \\ 2 & -5 \\ -1 & 0 \end{pmatrix}.$$

Notice that the transpose of a row vector is a column vector, and vice versa. Also, for any matrix $\mathbf{A}$ we have

$$(\mathbf{A}^{\mathrm{T}})^{\mathrm{T}} = \mathbf{A}.$$

Addition

Let $\mathbf{A}$ and $\mathbf{B}$ be two $m \times n$ matrices, say

$$\mathbf{A} = \begin{pmatrix} a_{11} & \cdots & a_{1n} \\ & \vdots & \\ a_{m1} & & a_{mn} \end{pmatrix}, \quad \mathbf{B} = \begin{pmatrix} b_{11} & \cdots & b_{1n} \\ & \vdots & \\ b_{m1} & \cdots & b_{mn} \end{pmatrix}.$$

We define $\mathbf{A} + \mathbf{B}$ to be the $m \times n$ matrix

$$\begin{pmatrix} a_{11} + b_{11} & \cdots & a_{1n} + b_{1n} \\ & \vdots & \\ a_{m1} + b_{m1} & \cdots & a_{mn} + b_{mn} \end{pmatrix}.$$

Thus addition of two matrices is defined when and only when the two matrices concerned have the same number of rows and the same number of columns, and is performed entry by entry. For example, if

$$\mathbf{A} = \begin{pmatrix} 1 & 2 & -1 \\ 4 & -5 & 0 \end{pmatrix} \quad \text{and} \quad \mathbf{B} = \begin{pmatrix} 3 & -4 & 2 \\ \frac{1}{2} & 1 & \sqrt{2} \end{pmatrix}$$

then

$$\mathbf{A} + \mathbf{B} = \begin{pmatrix} 4 & -2 & 1 \\ \frac{9}{2} & -4 & \sqrt{2} \end{pmatrix}.$$

It is easy to verify that for any two $m \times n$ matrices $\mathbf{A}$ and $\mathbf{B}$,

$$\mathbf{A} + \mathbf{B} = \mathbf{B} + \mathbf{A} \quad \text{and} \quad (\mathbf{A} + \mathbf{B})^{\mathrm{T}} = \mathbf{A}^{\mathrm{T}} + \mathbf{B}^{\mathrm{T}}.$$

The former identity is known as the *commutative property* of matrix addition. Also, if $\mathbf{A}$, $\mathbf{B}$ and $\mathbf{C}$ are three $m \times n$ matrices, then

$$(\mathbf{A} + \mathbf{B}) + \mathbf{C} = \mathbf{A} + (\mathbf{B} + \mathbf{C}). \tag{3}$$

Here we have another associative law, that for matrix addition: either side of (3) can be written without ambiguity as $\mathbf{A} + \mathbf{B} + \mathbf{C}$. Similarly, we may add more than three matrices without worrying about the order of summation.

The $m \times n$ matrix all of whose entries are zero is called the $m \times n$ *zero matrix* and is denoted by $\mathbf{O}_{mn}$. Thus $\mathbf{A} + \mathbf{O}_{mn} = \mathbf{A}$ for every $m \times n$ matrix $\mathbf{A}$. If there is no likelihood of confusion we write $\mathbf{O}_{mn}$ simply as $\mathbf{O}$.

Multiplication of a matrix by a scalar

Let $\mathbf{A}$ be an $m \times n$ matrix with (i, j) entry a_{ij} for $i = 1, \ldots, m$ and $j = 1, \ldots, n$. Let λ be any real number. Then we define $\lambda\mathbf{A}$ to be that $m \times n$ matrix whose (i, j) entry is λa_{ij} for $i = 1, \ldots, m$ and $j = 1, \ldots, n$. For example

$$\text{if } \mathbf{A} = \begin{pmatrix} 1 & 2 & -1 \\ 4 & -5 & 0 \end{pmatrix} \quad \text{then} \quad 3\mathbf{A} = \begin{pmatrix} 3 & 6 & -3 \\ 12 & -15 & 0 \end{pmatrix}.$$

The following general properties are easily demonstrated:

$$\lambda(\mathbf{A} + \mathbf{B}) = \lambda\mathbf{A} + \lambda\mathbf{B},$$
$$(\lambda\mathbf{A})^{\mathrm{T}} = \lambda\mathbf{A}^{\mathrm{T}}.$$

Subtraction

Given an $m \times n$ matrix $\mathbf{A}$, we denote the matrix $(-1)\mathbf{A}$ by $-\mathbf{A}$. Given two $m \times n$ matrices $\mathbf{A}$ and $\mathbf{B}$, we write

$$\mathbf{A} - \mathbf{B} = \mathbf{A} + (-\mathbf{B}).$$

Multiplication of matrices

Suppose we are given the $m \times n$ matrix

$$\mathbf{A} = \begin{pmatrix} a_{11} & \cdots & a_{1n} \\ & \vdots & \\ a_{m1} & \cdots & a_{mn} \end{pmatrix}$$

and the $n \times r$ matrix

$$\mathbf{B} = \begin{pmatrix} b_{11} & \cdots & b_{1r} \\ & \vdots & \\ b_{n1} & \cdots & b_{nr} \end{pmatrix}.$$

We then define the *matrix product* $\mathbf{AB}$ to be the $m \times r$ matrix whose (i, k) entry is

$$a_{i1}b_{1k} + a_{i2}b_{2k} + \ + \cdots + a_{in}b_{nk}$$

for $i = 1, \ldots, m$ and $k = 1, \ldots, r$.

Observe that the product $\mathbf{AB}$ is defined when and only when the number of columns of $\mathbf{A}$ is the same as the number of rows of $\mathbf{B}$. In this case, **AB** has the same number of rows as **A** and the same number of columns as **B**.

This notion of matrix multiplication is less complicated and more logical than might at first appear. To see its rationale, consider the case where $r = 1$. Then **B** is a column vector of length n, say

$$\mathbf{B} = \mathbf{x} = \begin{pmatrix} x_1 \\ \vdots \\ x_n \end{pmatrix}.$$

It then follows from the definition of matrix product that

$$\mathbf{Ax} = \begin{pmatrix} a_{11}x_1 + a_{12}x_2 + & \cdots & + a_{1n}x_n \\ a_{21}x_1 + a_{22}x_2 + & \cdots & + a_{2n}x_n \\ & \vdots & \\ a_{m1}x_1 + a_{m2}x_2 + & \cdots & + a_{mn}x_n \end{pmatrix}.$$

Thus for the case where $r = 1$, the justification of the way in which we defined matrix products is simply that it gives us our streamlined notation for systems of linear equations. The system (1) may be written in condensed form as

$$\mathbf{Ax} = \mathbf{b}$$

where

A is the $m \times n$ matrix given in (2),
x is the column vector $(x_1 \cdots x_n)^{\mathrm{T}}$ of length n
and **b** is the column vector $(b_1 \cdots b_m)^{\mathrm{T}}$ of length m.

Example 2 Let

$$\mathbf{A} = \begin{pmatrix} 4 & 1 & -2 \\ 0 & 2 & 1 \\ 1 & 0 & 0 \\ 1 & 2 & -3 \end{pmatrix}, \quad \mathbf{x} = \begin{pmatrix} 2 \\ 1 \\ 5 \end{pmatrix}, \quad \mathbf{w} = \begin{pmatrix} 0 \\ 2 \\ -1 \end{pmatrix}.$$

A is a 4×3 matrix and **x** and **w** are column vectors of length 3. Hence **Ax** and **Aw** are column vectors of length 4. We now compute these vectors.

$$\mathbf{Ax} = \begin{pmatrix} 4 \times 2 + 1 \times 1 - 2 \times 5 \\ 0 \times 2 + 2 \times 1 + 1 \times 5 \\ 1 \times 2 + 0 \times 1 + 0 \times 5 \\ 1 \times 2 + 2 \times 1 - 3 \times 5 \end{pmatrix} = \begin{pmatrix} -1 \\ 7 \\ 2 \\ -11 \end{pmatrix},$$

$$\mathbf{Aw} = \begin{pmatrix} 4 \times 0 + 1 \times 2 + 2 \times 1 \\ 0 \times 0 + 2 \times 2 - 1 \times 1 \\ 1 \times 0 + 0 \times 2 - 0 \times 1 \\ 1 \times 0 + 2 \times 2 + 3 \times 1 \end{pmatrix} = \begin{pmatrix} 4 \\ 3 \\ 0 \\ 7 \end{pmatrix}.$$

We have now defined the matrix product **AB**, and motivated this definition in the case where **B** has only one column. The passage to the general case is very simple: if the matrix product **AB** is defined and **B** has r columns, then:

$$(k\text{th column of } \mathbf{AB}) = \mathbf{A}(k\text{th column of } \mathbf{B})$$

for $k = 1, \ldots, r$. Also notice that

$$(i\text{th row of } \mathbf{AB}) = (i\text{th row of } \mathbf{A})\mathbf{B}$$

for $i = 1, \ldots, m$.

Example 3 Let **A** be the 4×3 matrix of Example 2 and let **B** be the 3×2 matrix

$$\begin{pmatrix} 2 & 0 \\ 1 & 2 \\ 5 & -1 \end{pmatrix}.$$

Letting the column vectors **x** and **w** be as in Example 2, we see that the first column of **B** is **x** and the second column of **B** is **w**. Thus **AB** is the 4×2 matrix whose first column is **Ax** and whose second column is **Aw**. Therefore

$$\mathbf{AB} = \begin{pmatrix} -1 & 4 \\ 7 & 3 \\ 2 & 0 \\ -11 & 7 \end{pmatrix}.$$

So much for the definition of matrix multiplication. The properties of this operation are sufficiently complicated to deserve a section to themselves.

Exercises

1. Given the matrices

$$\mathbf{A} = \begin{pmatrix} 2 & -1 & 0 \\ 1 & 3 & 5 \end{pmatrix}, \quad \mathbf{B} = \begin{pmatrix} 4 & 1 & 7 \\ -2 & 4 & 1 \end{pmatrix}, \quad \mathbf{C} = \begin{pmatrix} 2 & 1 & 3 \\ -4 & -5 & 8 \end{pmatrix},$$

find $\mathbf{A} + \mathbf{B}$, $3\mathbf{B} - 2\mathbf{C}$, $\mathbf{A} + 4\mathbf{B} - 2\mathbf{C}$ and $\mathbf{C}^{\mathrm{T}}$.

2. Given the matrices

$$\mathbf{A}=\begin{pmatrix}1&2&1\\2&0&-1\\1&-2&-2\\5&-2&-2\end{pmatrix},\quad \mathbf{B}=\begin{pmatrix}1&0\\2&1\\3&1\end{pmatrix},$$

$$\mathbf{C}=\begin{pmatrix}4&-4&3&1\\1&0&-1&0\end{pmatrix}$$

find **AB** and **BC**.

1.3 More on multiplication

Perhaps the most interesting fact about matrix multiplication is that it is *non-commutative*:

in general, $\mathbf{AB} \neq \mathbf{BA}$.

To see what this means, let **A** be $m \times n$ and **B** be $n \times r$. If $m \neq r$ then **BA** is not defined.

Now suppose that $m = r$. Then **AB** is a square matrix with m rows and **BA** is a square matrix with n rows. So if $m \neq n$ it is ungrammatical even to ask whether $\mathbf{AB} = \mathbf{BA}$.

The interesting case is where $m = n = r$. Here **A** is a square matrix with n rows, and so is **B**. The technical term for this is that **A** and **B** are square matrices of *order n*. Here **AB** and **BA** are also square matrices of order n, and they *may or may not* be the same matrix. For example:

$$\text{if } \mathbf{A}=\begin{pmatrix}2&0\\0&2\end{pmatrix} \quad\text{and}\quad \mathbf{B}=\begin{pmatrix}3&0\\0&3\end{pmatrix}$$

$$\text{then } \mathbf{AB}=\mathbf{BA}=\begin{pmatrix}6&0\\0&6\end{pmatrix};$$

BUT

$$\text{if } \mathbf{A}=\begin{pmatrix}0&1\\1&0\end{pmatrix} \quad\text{and}\quad \mathbf{B}=\begin{pmatrix}0&1\\-1&0\end{pmatrix}$$

$$\text{then } \mathbf{AB}=\begin{pmatrix}-1&0\\0&1\end{pmatrix} \quad\text{and}\quad \mathbf{BA}=\begin{pmatrix}1&0\\0&-1\end{pmatrix}.$$

The upshot of all this is that we have to mind our language when talking about matrix multiplication. Given two matrices **A** and **B**, the command 'multiply **A** by **B**' is ambiguous and should therefore never be used. For this reason we adopt the rather ugly words 'premultiply' and 'postmultiply'. Formation of the product **AB** is known as *premultiplication* of **B** by **A** and as *postmultiplication* of **A** by **B**. Similarly, we can refer to the formation of the product **BA** either as premultiplication of **A** by **B** or as postmultiplication of **B** by **A**.

Given that matrix multiplication is non-commutative, let us see what properties it does possess. First and foremost, it obeys an *associative law*, which has the effect of making expressions such as **ABCDE** unambiguous. Of all the results in this book, this is probably the one we shall appeal to most often:

$$(\mathbf{AB})\mathbf{C} = \mathbf{A}(\mathbf{BC}) \tag{1}$$

To prove (1), we must show that the typical entry of the matrix on the left is equal to the corresponding entry of the matrix on the right. In fact,

$$\sum_{k=1}^{r}\left(\sum_{j=1}^{n} a_{ij}b_{jk}\right)c_{kl} = \sum_{j=1}^{n}\sum_{k=1}^{r}(a_{ij}b_{jk}c_{kl}) = \sum_{j=1}^{n} a_{ij}\left(\sum_{k=1}^{r} b_{jk}c_{kl}\right)$$

so (1) is proved.

We now show how matrix multiplication interacts with multiplication by scalars, with addition and with transposition. The rules are

$$(\lambda\mathbf{A})\mathbf{B} = \lambda(\mathbf{AB}) = \mathbf{A}(\lambda\mathbf{B}), \tag{2}$$

$$(\mathbf{A} + \mathbf{G})\mathbf{B} = \mathbf{AB} + \mathbf{GB}, \tag{3}$$

$$\mathbf{A}(\mathbf{B} + \mathbf{H}) = \mathbf{AB} + \mathbf{AH}, \tag{4}$$

$$(\mathbf{AB})^{\mathrm{T}} = \mathbf{B}^{\mathrm{T}}\mathbf{A}^{\mathrm{T}}. \tag{5}$$

The identities (2), (3) and (4) can be proved in a manner similar to (1) by equating typical entries (but see Exercise 1 for a short cut). Identities (3) and (4) are known as the *distributive laws* of matrix addition and multiplication. Notice the reversion of order of multiplication in (5), which is proved as follows:

$$\begin{aligned}(k, i)\text{ entry of }\mathbf{B}^{\mathrm{T}}\mathbf{A}^{\mathrm{T}} &= \sum_{j=1}^{n} b_{jk}a_{ij}\\ &= \sum_{j=1}^{n} a_{ij}b_{jk}\\ &= (i, k)\text{ entry of }\mathbf{AB}\\ &= (k, i)\text{ entry of }(\mathbf{AB})^{\mathrm{T}}.\end{aligned}$$

By combining (1)–(5) we can perform all sorts of operations. For example,

$$(3\mathbf{A} + 2\mathbf{G})(\mathbf{BC} - \mathbf{HC}) = (3\mathbf{AB} + 2\mathbf{GB} - 3\mathbf{AH} - 2\mathbf{GH})\mathbf{C},$$
$$\mathbf{A}(4\mathbf{B} - 5\mathbf{H}) = 4\mathbf{AB} - 5\mathbf{AH},$$
$$2\mathbf{B}^{\mathrm{T}}\mathbf{A}^{\mathrm{T}} + 3\mathbf{B}^{\mathrm{T}}\mathbf{G}^{\mathrm{T}} = ((2\mathbf{A} + 3\mathbf{G})\mathbf{B})^{\mathrm{T}}.$$

In the preceding section, we discussed not only addition of matrices but also subtraction. We also mentioned zero matrices, which play the same rôle in matrix algebra that the number zero plays in ordinary arithmetic. Can we accompany our discussion of matrix multiplication with a description of matrix division and matrix analogies to the number one?

There is no such thing as matrix division. The closest analogy to it will be discussed in detail in Section 2.2.

On the other hand we can say something now about matrix analogies to the number 1 (unity). Recall that a square matrix of order n is an $n \times n$ matrix. Let $\mathbf{A} = (a_{ij})$ be such a matrix. Then

$$a_{11}, a_{22}, \ldots, a_{nn}$$

are said to be the *diagonal entries* of **A**, and all other entries are said to be *offdiagonal*. The *identity matrix* (or *unit matrix*) of order n is that $n \times n$ matrix whose diagonal entries are all 1 and whose offdiagonal entries are all 0. The identity matrix of order n will be denoted by $\mathbf{I}_n$, or simply by **I** when there is no ambiguity about n. Thus

$$\mathbf{I}_2 = \begin{pmatrix} 1 & 0 \\ 0 & 1 \end{pmatrix}, \quad \mathbf{I}_3 = \begin{pmatrix} 1 & 0 & 0 \\ 0 & 1 & 0 \\ 0 & 0 & 1 \end{pmatrix}$$

and so on.

Another way of phrasing the definition of an identity matrix is to say that it has 1's 'down the diagonal' and 0's elsewhere. Also 'diagonal' when used as a noun is sometimes qualified by the adjective 'main' but we shall not use the qualifier in this book.

Let us now see in what sense the rôle played by unit matrices in matrix multiplication is analogous to the rôle of unity in ordinary arithmetic. It is easy to show that if **x** is any column vector of length n, then

$$\mathbf{I}_n\mathbf{x} = \mathbf{x}.$$

Simlarly, if **p** is a row vector of length n, then

$$\mathbf{p}\mathbf{I}_n = \mathbf{p}.$$

It follows that if $\mathbf{A}$ is any $m \times n$ matrix, then

$$\mathbf{I}_m \mathbf{A} = \mathbf{A}\mathbf{I}_n = \mathbf{A}.$$

We end this section with a few remarks about *partitioned matrices*. These are most easily illustrated by an example. Let $\mathbf{A}$ be the 3×6 matrix

$$\begin{pmatrix} 1 & 5 & 7 & 8 & 2 & 0 \\ 3 & -1 & 4 & 0 & 0 & 1 \\ -1 & 0 & 2 & 1 & 0 & -1 \end{pmatrix}.$$

Then we can write

$$\mathbf{A} = \begin{pmatrix} \mathbf{A}_1 & \mathbf{A}_2 & \mathbf{A}_3 \\ \mathbf{A}_4 & \mathbf{A}_5 & \mathbf{A}_6 \end{pmatrix}$$

where

$$\mathbf{A}_1 = \begin{pmatrix} 1 & 5 \\ 3 & -1 \end{pmatrix}, \quad \mathbf{A}_2 = \begin{pmatrix} 7 & 8 & 2 \\ 4 & 0 & 0 \end{pmatrix}, \quad \mathbf{A}_3 = \begin{pmatrix} 0 \\ 1 \end{pmatrix},$$

$$\mathbf{A}_4 = (-1 \quad 0), \quad \mathbf{A}_5 = (2 \quad 1 \quad 0), \quad \mathbf{A}_6 = (-1).$$

When we break up $\mathbf{A}$ in such a way, we are said to *partition* $\mathbf{A}$ into the *blocks* $\mathbf{A}_1, \ldots, \mathbf{A}_6$. Notice that the partition we have just described is not the only one possible. For example, we can also write the same matrix $\mathbf{A}$ as

$$\begin{pmatrix} \mathbf{B}_1 & \mathbf{B}_2 \\ \mathbf{B}_3 & \mathbf{B}_4 \end{pmatrix}$$

where

$$\mathbf{B}_1 = (1 \quad 5 \quad 7), \quad \mathbf{B}_2 = (8 \quad 2 \quad 0),$$

$$\mathbf{B}_3 = \begin{pmatrix} 3 & -1 & 4 \\ -1 & 0 & 2 \end{pmatrix}, \quad \mathbf{B}_4 = \begin{pmatrix} 0 & 0 & 1 \\ 1 & 0 & -1 \end{pmatrix}.$$

The most important fact about partitioned matrices concerns multiplication. Suppose that we wish to premultiply a matrix which is partitioned into blocks by another matrix, also partitioned into blocks: then provided the relevant block-products are defined, we can use the usual rule of matrix multiplication, *applied to blocks rather than to individual entries*. For example, suppose that

$$\mathbf{G} = \begin{pmatrix} \mathbf{G}_1 & \mathbf{G}_2 \\ \mathbf{G}_3 & \mathbf{G}_4 \end{pmatrix}, \quad \mathbf{H} = \begin{pmatrix} \mathbf{H}_1 & \mathbf{H}_2 \\ \mathbf{H}_3 & \mathbf{H}_4 \end{pmatrix},$$

where the number of columns of $\mathbf{G}$ equals the number of columns of $\mathbf{H}$, and the number of columns of $\mathbf{G}_1$ equals the number of rows of $\mathbf{H}_1$. Then

$$\mathbf{GH} = \begin{pmatrix} \mathbf{G}_1\mathbf{H}_1 + \mathbf{G}_2\mathbf{H}_3 & \mathbf{G}_1\mathbf{H}_2 + \mathbf{G}_2\mathbf{H}_4 \\ \mathbf{G}_3\mathbf{H}_1 + \mathbf{G}_4\mathbf{H}_3 & \mathbf{G}_3\mathbf{H}_2 + \mathbf{G}_4\mathbf{H}_4 \end{pmatrix}.$$

The proof of this fact is omitted on grounds of tedium: the reader is encouraged to verify it with numerical examples.

Exercises

1. Derive the distributive law (3) from the distributive law (4) and the transposition rule (5).

2. Given

$$\mathbf{A} = \begin{pmatrix} 3 & 1 & 2 \\ 4 & 1 & 0 \end{pmatrix}, \quad \mathbf{B} = \begin{pmatrix} 5 & 1 & 9 \\ -1 & 2 & 5 \end{pmatrix}, \quad \mathbf{C} = \begin{pmatrix} -2 & 4 \\ -3 & 6 \end{pmatrix},$$

find $3\mathbf{A} - 2\mathbf{B}$, $2\mathbf{C}^T\mathbf{B} - 3\mathbf{C}^T\mathbf{A}$, $6\mathbf{A}^T\mathbf{C} - 4\mathbf{B}^T\mathbf{C}$ and $\mathbf{C}^T(2\mathbf{B} - 3\mathbf{A})\mathbf{C}$.

3. The *square* of a square matrix $\mathbf{A}$ is the matrix $\mathbf{A}^2 = \mathbf{AA}$.
   (i) Show that

   $$(\mathbf{A} + \mathbf{B})^2 = \mathbf{A}^2 + \mathbf{B}^2 + \mathbf{BA} + \mathbf{AB}$$

   and that $\mathbf{A}^2 - \mathbf{B}^2$ is not in general equal to $(\mathbf{A} + \mathbf{B})(\mathbf{A} - \mathbf{B})$.
   (ii) Find all 2×2 matrices $\mathbf{A}$ for which $\mathbf{A}^2 = \mathbf{O}$.
   (iii) Find all 2×2 matrices $\mathbf{A}$ for which $\mathbf{A}^2 = -\mathbf{I}$.

1.4 Vectors

Up to this point, vectors have been considered simply as a special case of matrices. We now look at them in more detail. *For the rest of this book, the term, 'vector' will always be used to mean COLUMN vector.* A column vector of length n will be referred to as an *n-vector*. The set of all n-vectors is referred to as *real n-space* and denoted by R^n.

Writing vectors as columns is convenient and conventional, but it can take up a lot of space. Let $\mathbf{x} \in R^n$, $\mathbf{y} \in R^m$. Then we can define the

$(n+m)$-vector

$$\mathbf{z} = \begin{pmatrix} \mathbf{x} \\ \mathbf{y} \end{pmatrix} = \begin{pmatrix} x_1 \\ \vdots \\ x_n \\ y_1 \\ \vdots \\ y_m \end{pmatrix}.$$

Clearly we can save a little space by writing $\mathbf{z} = (x_1 \cdots x_n \; y_1 \cdots y_m)^{\mathrm{T}}$. We can save even more with the notation $\mathbf{z} = (\mathbf{x} : \mathbf{y})$ and this we shall adopt. Expressions such as $(\mathbf{x} : \mathbf{y} : \mathbf{a} : \mathbf{b})$ have similar meanings.

The zero n-vector is that member of R^n whose components are all zero and we denote it by $\mathbf{0}$.

The first major fact about vectors concerns matrix equality and is simple and crucial.

THEOREM 1 Given $m \times n$ matrices $\mathbf{A}$ and $\mathbf{B}$, we have $\mathbf{A} = \mathbf{B}$ if and only if $\mathbf{Ax} = \mathbf{Bx}$ for every n-vector $\mathbf{x}$.

PROOF 'Only if' is obvious. To prove 'if' let $\mathbf{Ax} = \mathbf{Bx}$ for all $\mathbf{x}$ in R^n. Then this is true in particular when $\mathbf{x}$ is any column of $\mathbf{I}_n$. Thus

$$\mathbf{A} = \mathbf{AI} = \mathbf{BI} = \mathbf{B}. \qquad \text{QED}$$

We now consider the standard way in which vectors are combined to produce other vectors. We say that the n-vector $\mathbf{b}_1$ is a *linear combination* of the n-vectors $\mathbf{b}_2, \ldots, \mathbf{b}_k$ if there exist scalars $\lambda_2, \ldots, \lambda_k$ such that

$$\mathbf{b}_1 = \lambda_2 \mathbf{b}_2 + \cdots + \lambda_k \mathbf{b}_k. \tag{1}$$

The first thing to notice about this is that $\mathbf{b}_1$ is a linear combination of $\mathbf{b}_2, \ldots, \mathbf{b}_k$ if and only if there exist scalars $\alpha_1, \alpha_2, \ldots, \alpha_k$ such that

$$\alpha_1 \mathbf{b}_1 + \alpha_2 \mathbf{b}_2 + \cdots + \alpha_k \mathbf{b}_k = \mathbf{0} \tag{2}$$

and $\alpha_1 \neq 0$. For if (1) is true then (2) holds with $\alpha_1 = -1$ and $\alpha_j = \lambda_j$ for $j = 2, \ldots, k$. Conversely, if (2) is true and $\alpha_1 \neq 0$ we may retrieve (1) by setting $\lambda_j = -\alpha_j/\alpha_1$ for $j = 2, \ldots, k$.

We say that the vectors $\mathbf{b}_1, \ldots, \mathbf{b}_k$ are *linearly dependent* if it is possible to express one of them as a linear combination of the others. Since there is nothing special about $\mathbf{b}_1$ we may generalise the result of the last paragraph as follows: $\mathbf{b}_1, \ldots, \mathbf{b}_k$ are linearly dependent if and only if (2) holds for some scalars $\alpha_1, \ldots, \alpha_k$ *which are not all zero.*

We say that $\mathbf{b}_1, \ldots, \mathbf{b}_k$ are *linearly independent* vectors if they are not linearly dependent. There are two equivalent ways of spelling this out in full: $\mathbf{b}_1, \ldots, \mathbf{b}_k$ are linearly independent if and only if it is not possible to express any one of them as a linear combination of the others; and $\mathbf{b}_1, \ldots, \mathbf{b}_k$ are linearly independent if and only if the only list of scalars $\alpha_1, \ldots, \alpha_k$ satisfying (2) is given by

$$\alpha_1 = \alpha_2 = \cdots = \alpha_k = 0.$$

If $\mathbf{b}_1, \ldots, \mathbf{b}_k$ are linearly independent they must be different from each other and from $\mathbf{0}$. For if $\mathbf{b}_1 = \mathbf{0}$ then

$$1\mathbf{b}_1 + 0\mathbf{b}_2 + \cdots + 0\mathbf{b}_k = \mathbf{0},$$

contradicting the hypothesis of linear independence. If $\mathbf{b}_1 = \mathbf{b}_2$ then

$$1\mathbf{b}_1 - 1\mathbf{b}_2 + 0\mathbf{b}_3 + \cdots + 0\mathbf{b}_k = \mathbf{0},$$

again contradicting linear independence.

Notice how the definitions of linear combination, linear dependence and linear independence work out in terms of matrices. Let $\mathbf{B}$ be an $n \times k$ matrix. The n-vector $\mathbf{y}$ is a linear combination of the columns of $\mathbf{B}$ if and only if $\mathbf{y} = \mathbf{Bz}$ for some k-vector $\mathbf{z}$. The columns of $\mathbf{B}$ are linearly dependent if and only if there exists a k-vector $\mathbf{w}$ such that

$$\mathbf{Bw} = \mathbf{0} \quad \text{and} \quad \mathbf{w} \neq \mathbf{0}.$$

The columns of $\mathbf{B}$ are linearly independent if and only if

$$\mathbf{Bx} = \mathbf{0} \text{ implies } \mathbf{x} = \mathbf{0}.$$

It is particularly important to observe that the columns of $\mathbf{I}_n$ are n linearly independent n-vectors. For

$$\text{if } \mathbf{Ix} = \mathbf{0} \text{ then } \mathbf{x} = \mathbf{Ix} = \mathbf{0}.$$

Example 1 Given

$$\mathbf{b}_1 = \begin{pmatrix} 1 \\ 8 \end{pmatrix}, \quad \mathbf{b}_2 = \begin{pmatrix} 1 \\ 2 \end{pmatrix}, \quad \mathbf{b}_3 = \begin{pmatrix} 2 \\ 1 \end{pmatrix},$$

we have $\mathbf{b}_1 = 5\mathbf{b}_2 - 2\mathbf{b}_3$ so $\mathbf{b}_1, \mathbf{b}_2, \mathbf{b}_3$ are linearly dependent. On the other hand, $\mathbf{b}_2$ and $\mathbf{b}_3$ are linearly independent: for

$$\text{if } \lambda\mathbf{b}_2 + \mu\mathbf{b}_3 = \mathbf{0} \text{ then } \lambda + 2\mu = 2\lambda + \mu = 0 \text{ so } \lambda = \mu = 0.$$

Putting this in matrix terms, let

$$\mathbf{A} = \begin{pmatrix} 1 & 2 \\ 2 & 1 \end{pmatrix}, \quad \mathbf{B} = \begin{pmatrix} 1 & 1 & 2 \\ 8 & 2 & 1 \end{pmatrix}.$$

Then the columns of **A** are linearly independent, while the columns of **B** are linearly dependent.

We now state a theorem which will play such an important part in the next chapter that we give it an imposing name. The Second Law will appear in Section 2.1.

THEOREM 2 (First Law of Linear Dependence)
Let $\mathbf{b}_1, \mathbf{b}_2, \ldots, \mathbf{b}_k$ be linearly dependent n-vectors such that $\mathbf{b}_2, \ldots, \mathbf{b}_k$ are linearly independent. Then $\mathbf{b}_1$ is a linear combination of $\mathbf{b}_2, \ldots, \mathbf{b}_k$.

PROOF By assumption, there exist scalars $\alpha_1, \alpha_2, \ldots, \alpha_k$, not all zero, such that

$$\alpha_1 \mathbf{b}_1 + \alpha_2 \mathbf{b}_2 + \cdots + \alpha_k \mathbf{b}_k = \mathbf{0}.$$

We want to show that $\alpha_1 \neq 0$.

If α_1 were zero we would have

$$\alpha_2 \mathbf{b}_2 + \cdots + \alpha_k \mathbf{b}_k = \mathbf{0} \quad \text{and} \quad a_j \neq 0 \text{ for some } j > 1$$

But this contradicts the linear independence of $\mathbf{b}_2, \ldots, \mathbf{b}_k$. QED

Given two n-vectors $\mathbf{x}$ and $\mathbf{y}$ we define the *inner product* of $\mathbf{x}$ and $\mathbf{y}$ to be the scalar

$$\mathbf{x} \cdot \mathbf{y} = x_1 y_1 + x_2 y_2 + \cdots + x_n y_n.$$

Evidently $\mathbf{x} \cdot \mathbf{y} = \mathbf{y} \cdot \mathbf{x}$. Also the 1×1 matrix $\mathbf{x}^{\mathrm{T}}\mathbf{y}$ is $(\mathbf{x} \cdot \mathbf{y})$. In the sequel we shall simply identify 1×1 matrices with scalars and drop the dot notation: the inner product of $\mathbf{x}$ and $\mathbf{y}$ is written $\mathbf{x}^{\mathrm{T}}\mathbf{y}(=\mathbf{y}^{\mathrm{T}}\mathbf{x})$. The phrases '$\mathbf{x}$ is *orthogonal* to $\mathbf{y}$', '$\mathbf{y}$ is orthogonal to $\mathbf{x}$' and '$\mathbf{x}$ and $\mathbf{y}$ are orthogonal' all mean the same thing, namely that $\mathbf{x}^{\mathrm{T}}\mathbf{y} = 0$. For example, if

$$\mathbf{x} = \begin{pmatrix} 2 \\ 3 \\ -1 \\ 0 \end{pmatrix}, \quad \mathbf{y} = \begin{pmatrix} 1 \\ 0 \\ 2 \\ 4 \end{pmatrix},$$

then $\mathbf{x}^T\mathbf{y} = 2 \times 1 + 3 \times 0 - 1 \times 2 + 0 \times 4 = 0$, so $\mathbf{x}$ and $\mathbf{y}$ are orthogonal.

The following theorem says what happens when we take the inner product of a vector with itself.

THEOREM 3 $\mathbf{x}^T\mathbf{x} \geq 0$, with equality if and only if $\mathbf{x} = \mathbf{0}$.

PROOF For each component j, $x_j x_j = x_j^2 \geq 0$. A sum of non-negative terms is non-negative, being zero if and only if each term is zero.

QED

In view of Theorem 3 we may define for each n-vector $\mathbf{x}$ the non-negative number

$$\|\mathbf{x}\| = \sqrt{\mathbf{x}^T\mathbf{x}} = \sqrt{x_1^2 + x_2^2 + \cdots + x_n^2}.$$

We call $\|\mathbf{x}\|$ the *Euclidean norm* of $\mathbf{x}$ and will usually omit the adjective: other norms exist, but they will not be mentioned in this book. Of course 'Euclidean' has geometrical connotations; so does 'orthogonal' (which means 'at right angles'); so for that matter does 'linear'. But the geometry of vectors will not be revealed until we have established the main properties of the Euclidean norm.

THEOREM 4
(i) $\|\mathbf{x}\| \geq 0$, with equality if and only if $\mathbf{x} = \mathbf{0}$
(ii) $\|\lambda\mathbf{x}\| = |\lambda| \cdot \|\mathbf{x}\|$
(iii) $|\mathbf{x}^T\mathbf{y}| \leq \|\mathbf{x}\| \cdot \|\mathbf{y}\|$
(iv) $\|\mathbf{x} + \mathbf{y}\| \leq \|\mathbf{x}\| + \|\mathbf{y}\|$

PROOF $\|\mathbf{x}\| \geq 0$ by definition, and the rest of (i) is Theorem 3. Also

$$\|\lambda\mathbf{x}\|^2 = (\lambda\mathbf{x})^T(\lambda\mathbf{x}) = \lambda^2\mathbf{x}^T\mathbf{x}.$$

Taking square roots we obtain (ii).

To prove (iii) and (iv) let

$$\alpha = \|\mathbf{x}\|, \qquad \beta = \|\mathbf{y}\|, \quad \theta = \mathbf{x}^T\mathbf{y}.$$

If $\mathbf{y} = \mathbf{0}$, both sides of (iii) are zero and both sides of (iv) are $\|\mathbf{x}\|$. So suppose from now on that $\mathbf{y} \neq \mathbf{0}$: then $\alpha \geq 0$ and $\beta > 0$.

For any scalar μ, $(\mathbf{x} - \mu\mathbf{y})^T(\mathbf{x} - \mu\mathbf{y}) = \mathbf{x}^T\mathbf{x} - 2\mu\mathbf{x}^T\mathbf{y} + \mu^2\mathbf{y}^T\mathbf{y}$. Thus

$$\|\mathbf{x} - \mu\mathbf{y}\|^2 = \alpha^2 - 2\mu\theta + \mu^2\beta^2. \tag{3}$$

By (3), $\alpha^2 \geq \mu(2\theta - \mu\beta^2)$ for any μ. Setting $\mu = \theta/\beta^2$ we have $\alpha^2 \geq (\theta/\beta)^2$. Taking square roots and multiplying by β we obtain (iii).

Setting $\mu = -1$ in (3) we have $\|\mathbf{x} + \mathbf{y}\|^2 = \alpha^2 + 2\theta + \beta^2$. Thus

$$(\alpha + \beta)^2 - \|\mathbf{x} + \mathbf{y}\|^2 = 2(\alpha\beta - \theta),$$

which is non-negative by (iii). Thus $\|\mathbf{x} + \mathbf{y}\|^2 \leq (\alpha + \beta)^2$. Taking square roots we obtain (iv). QED

Part (iii) of Theorem 4 is known in anglophone countries as the *Schwarz inequality*: the French call it Cauchy's inequality and the Russians Bunyakovskii's inequality. (For biographical information see Bartle, 1976, p. 56.) Part (iv) is known as the *triangle inequality*, for reasons that will shortly become apparent.

Consider 2-space. Given any two 2-vectors

$$\mathbf{a} = \begin{pmatrix} a_1 \\ a_2 \end{pmatrix} \quad \text{and} \quad \mathbf{b} = \begin{pmatrix} b_1 \\ b_2 \end{pmatrix},$$

we can represent them as points in the plane, relative to two given coordinate axes. The vector $\mathbf{a}$ is represented by the point A = (a_1, a_2) and $\mathbf{b}$ by B = (b_1, b_2). This is illustrated in Figure 1.1 for the case where a_1, a_2 and b_1 are positive and b_2 is negative.

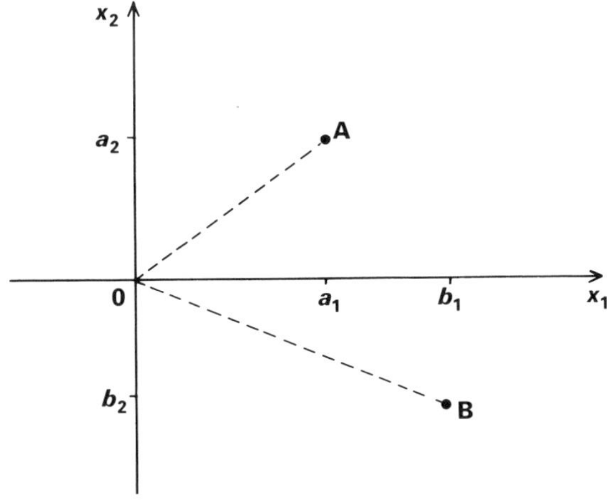

FIG. 1.1

As the reader may know, the distance between the points A and B is

$$\sqrt{(a_1 - b_1)^2 + (a_2 - b_2)^2} = \|\mathbf{a} - \mathbf{b}\|.$$

Similarly, $\|\mathbf{a}\|$ is the length of the line OA and $\|\mathbf{b}\|$ the length of OB. Now

$$\overline{\text{OA}}^2 + \overline{\text{OB}}^2 - \overline{\text{AB}}^2 = \mathbf{a}^{\text{T}}\mathbf{a} + \mathbf{b}^{\text{T}}\mathbf{b} - (\mathbf{a} - \mathbf{b})^{\text{T}}(\mathbf{a} - \mathbf{b}) = 2\mathbf{a}^{\text{T}}\mathbf{b}.$$

So by Pythagoras' Theorem, $\mathbf{a}^{\text{T}}\mathbf{b} = 0$ if and only if the line OA is at right angles (orthogonal) to the line OB.

We can carry this definition of distance over to n-space. Given n-vectors $\mathbf{a}$ and $\mathbf{b}$, we define the (Euclidean) *distance* between them to be

$$\text{dis}\,(\mathbf{a}, \mathbf{b}) = \|\mathbf{a} - \mathbf{b}\|.$$

It is important to notice that this definition has the three properties associated with the intuitive notion of distance:

(α) dis $(\mathbf{a}, \mathbf{b}) \geq 0$, with equality if and only if $\mathbf{a} = \mathbf{b}$;

(β) dis $(\mathbf{a}, \mathbf{b}) =$ dis $(\mathbf{b}, \mathbf{a})$;

(γ) dis $(\mathbf{a}, \mathbf{b}) +$ dis $(\mathbf{b}, \mathbf{c}) \geq$ dis $(\mathbf{a}, \mathbf{c})$.

These are all consequences of Theorem 4: (α) follows from (i) with $\mathbf{x} = \mathbf{a} - \mathbf{b}$, ($\beta$) from (ii) with $\mathbf{x} = \mathbf{a} - \mathbf{b}$ and $\lambda = -1$ and (γ) from (iv) with $\mathbf{x} = \mathbf{a} - \mathbf{b}$ and $\mathbf{y} = \mathbf{b} - \mathbf{c}$.

We say that (γ) is part of the intuitive notion of distance because it captures the idea that the length of one side of a triangle cannot exceed the sum of the lengths of the other two sides. It should now be fairly obvious why (iv) is called the triangle inequality. We say *fairly* obvious because the elementary geometry of the plane yields a stronger result than the one just stated: the length of one side of a triangle is *less* than the sum of the lengths of the other two sides, except in the degenerate case where the triangle collapses into a line.

Before seeing how this works out in n-space, let us get some facts straight about line segments. If $\text{A} = (a_1, a_2)$ and $\text{B} = (b_1, b_2)$ are points in the plane, the midpoint of the line AB is the point $(\frac{1}{2}a_1 + \frac{1}{2}b_1, \frac{1}{2}a_2 + \frac{1}{2}b_2)$. Similarly, to split AB into three lines of equal length is to mark the points

$$\left(\frac{2}{3}a_1 + \frac{1}{3}b_1, \frac{2}{3}a_2 + \frac{1}{3}b_2\right) \quad \text{and} \quad \left(\frac{1}{3}a_1 + \frac{2}{3}b_1, \frac{1}{3}a_2 + \frac{2}{3}b_2\right).$$

Generally we can consider the line AB as being traced out by all points of the form $((1-\theta)a_1+\theta b_1, (1-\theta)a_2+\theta b_2))$ as the variable θ increases from 0 to 1. This idea can be applied directly to R^n: the *line segment* joining n-vectors $\mathbf{a}$ and $\mathbf{b}$ is defined as the set of all vectors of the form $(1-\theta)\mathbf{a}+\theta\mathbf{b}$, where $0 \le \theta \le 1$.

Now back to triangles. The reader is asked to show in Exercise 5 that $\|\mathbf{x}+\mathbf{y}\| = \|\mathbf{x}\| + \|\mathbf{y}\|$ if and only if $\lambda\mathbf{x} = \mu\mathbf{y}$ for some non-negative scalars λ and μ which are not both zero. Given this special kind of linear dependence we can set $\theta = \mu/(\lambda+\mu)$, inferring that

$$0 \le \theta \le 1 \quad \text{and} \quad (1-\theta)\mathbf{x}+\theta(-\mathbf{y}) = \mathbf{0}.$$

Thus $\|\mathbf{x}+\mathbf{y}\| = \|\mathbf{x}\| + \|\mathbf{y}\|$ if and only if $\mathbf{0}$ is on the line segment joining $\mathbf{x}$ and $(-\mathbf{y})$. Setting $\mathbf{x} = \mathbf{a}-\mathbf{b}$ and $\mathbf{y} = \mathbf{b}-\mathbf{c}$ we obtain the strong form of the triangle inequality in R^n:

$$\|\mathbf{a}-\mathbf{c}\| \le \|\mathbf{a}-\mathbf{b}\| + \|\mathbf{b}-\mathbf{c}\|$$

with equality if and only if $\mathbf{b}$ is on the line segment joining $\mathbf{a}$ and $\mathbf{c}$.

Exercises

1.

Given $\mathbf{A} = \begin{pmatrix} 1 & 1 & 2 & 1 & 3 \\ 1 & 0 & 1 & 2 & 3 \\ 1 & 1 & 0 & 1 & 1 \end{pmatrix}$ prove:

(i) columns 1, 2, 3 of $\mathbf{A}$ are linearly independent;
(ii) columns 1, 2, 4 of $\mathbf{A}$ are linearly dependent;
(iii) columns 3, 4, 5 of $\mathbf{A}$ are linearly dependent

2. Prove that if $\mathbf{x}$, $\mathbf{y}$, $\mathbf{z}$ are non-zero vectors such that $\mathbf{y}^T\mathbf{z} = \mathbf{z}^T\mathbf{x} = \mathbf{x}^T\mathbf{y} = 0$, then $\mathbf{x}$, $\mathbf{y}$, $\mathbf{z}$ are linearly independent.

3. Show that $\|3\mathbf{x}-2\mathbf{y}\| \ge 3\|\mathbf{x}\| - 2\|\mathbf{y}\|$ and that

$$\|3\mathbf{x}-2\mathbf{y}+6\mathbf{z}\| \le 3\|\mathbf{x}\| + 2\|\mathbf{y}\| + 6\|\mathbf{z}\|.$$

4. Draw diagrams illustrating the following sets in R^2.
(i) $\{\mathbf{x} \in R^2 \colon \|\mathbf{x}\| \le 3\}$
(ii) $\{\mathbf{x} \in R^2 \colon x_2 \le x_1^2\}$
(iii) $\{\mathbf{x} \in R^2 \colon x_1 \ge 3 \text{ and } x_2 \ge 1 + x_1\}$

5. The strong form of the Schwarz inequality states that $|\mathbf{x}^{\mathrm{T}}\mathbf{y}| \leq \|\mathbf{x}\| \cdot \|\mathbf{y}\|$, with equality if and only if $\mathbf{x}$ and $\mathbf{y}$ are linearly dependent. Prove this. Hence show that

$$\|\mathbf{x} + \mathbf{y}\| \leq \|\mathbf{x}\| + \|\mathbf{y}\|$$

with equality if and only if there exist *non-negative* scalars λ and μ, not both zero, such that $\lambda\mathbf{x} = \mu\mathbf{y}$.

2 Linear Equations

In this chapter we explain how to solve a system of m linear equations in n unknowns. It might seem natural to start with the case $m = n$. This we do, but we also find it convenient to impose a restrictive property on the matrix of coefficients. This property is called nonsingularity and we say more about it in Section 2.2. Section 2.3 sets up the equipment for handling the general $m \times n$ case, and all is revealed in Section 2.4.

2.1 Gaussian elimination

We begin by defining some special kinds of square matrix. A square matrix is said to be *upper triangular* if it has only zeroes *below* the diagonal, *lower triangular* if it has only zeroes *above* the diagonal, *triangular* if it is either upper or lower triangular and *diagonal* if it is both. If

$$\mathbf{U} = \begin{pmatrix} 1 & 3 & 2 \\ 0 & 2 & 1 \\ 0 & 0 & 5 \end{pmatrix}, \quad \mathbf{L} = \begin{pmatrix} 3 & 0 & 0 \\ 4 & 0 & 0 \\ 5 & 1 & 2 \end{pmatrix}, \quad \mathbf{D} = \begin{pmatrix} 1 & 0 & 0 \\ 0 & 3 & 0 \\ 0 & 0 & 4 \end{pmatrix},$$

then $\mathbf{U}$ is upper triangular, $\mathbf{L}$ is lower triangular and $\mathbf{D}$ is diagonal. The identity matrix $\mathbf{I}_3$ is another 3×3 diagonal matrix.

The $n \times n$ equation system

$$\mathbf{Ax} = \mathbf{b}$$

is said to be a *Nonsingular Upper Triangular* system (NUT, for short) if the coefficient matrix $\mathbf{A}$ is upper triangular and none of its diagonal entries is zero. The force of the term 'nonsingular' will become clear soon.

The point about NUTs is that they can be solved very simply by a process called *back-substitution.*

Example 1 Solve the system of linear equations

$$\begin{aligned} 3x_1 - 4x_2 - 2x_3 &= 4 \\ 3x_2 + x_3 &= 5 \\ 4x_3 &= 8. \end{aligned}$$

We solve from the bottom up. Clearly $x_3 = 2$, whence $x_2 = (5 - 2)/3 = 1$, whence $x_1 = (4 + 4 + 4)/3 = 4$.

A square matrix is said to be *singular* if its columns are linearly dependent, *nonsingular* if its columns are linearly independent. Notice that the idea of linear dependence or independence of columns is applicable whether or not the matrix is square, but the terms 'singular' and 'nonsingular' are used *only* for square matrices.

The facts about linear dependence given in the last section enable us to recognise some singular and nonsingular matrices. For example, a square matrix with a column consisting entirely of zeros is singular, as is a square matrix with two identical columns. Also, $\mathbf{I}_n$ is a *nonsingular* matrix. This fact has the important consequence that there exist nonsingular matrices of every order. It also has the following generalisation, which explains the N in NUT.

THEOREM 1 A triangular matrix is singular if and only if at least one of its diagonal entries is zero.

PROOF Let the $n \times n$ matrix $\mathbf{A}$ be upper triangular. Suppose we try to solve the system $\mathbf{Ax} = \mathbf{0}$ by back-substitution. If no diagonal entry is zero, we obtain after n steps the unique solution $\mathbf{x} = \mathbf{0}$: hence $\mathbf{A}$ is nonsingular. If $a_{kk} = 0$ for some k, we can put $x_k = 1$ at step $(n + 1 - k)$ of the substitution process and continue to the next step. This procedure, which is illustrated in Example 2, yields a non-zero $\mathbf{x}$ for which $\mathbf{Ax} = \mathbf{0}$: thus $\mathbf{A}$ is singular.

The lower triangular case of the theorem is proved similarly, working from the top down. QED

Example 2 Find a non-trivial solution to the system

$$\begin{aligned} 3x_1 - 6x_2 - 3x_3 + 5x_4 &= 0 \\ x_3 + 2x_4 &= 0 \\ x_3 - x_4 &= 0 \\ x_4 &= 0 \end{aligned}$$

This is a singular upper triangular system: the (2, 2) entry of the coefficient matrix is zero. We start by solving $x_4 = 0$ and then $x_3 = 0$. At Step $3(= 4 + 1 - 2)$ of the process, let $x_2 = 1$. Then the first equation implies that $x_1 = 2$. We have as a non-trivial solution

$$x_1 = 2, \quad x_2 = 1, \quad x_3 = 0, \quad x_4 = 0.$$

We say that the system $\mathbf{Ax} = \mathbf{b}$ is singular if $\mathbf{A}$ is singular, nonsingular otherwise. We shall prove by induction a theorem about nonsingular systems: we then show how the logic of the proof yields *the* practical method of solving such systems.

THEOREM 2 Let $\mathbf{A}$ be a nonsingular matrix of order n, $\mathbf{b}$ an n-vector. Then the system $\mathbf{Ax} = \mathbf{b}$ has a unique solution.

PROOF We proceed by induction on n. For $n = 1$, the result is ordinary division: notice that it is the assumption of nonsingularity which prevents division by zero. Now let $n > 1$, and suppose inductively that the required result holds for systems of order $n - 1$. Since $\mathbf{A}$ is nonsingular, the first column of $\mathbf{A}$ cannot be a zero vector: there exists some k for which $a_{k1} \neq 0$. Now the equations of the system can be written in any order without changing the system. We therefore assume without loss of generality that $a_{11} \neq 0$. (The reader is encouraged to be wary of sentences including the phrase 'without loss of generality' and we shall comment on the last one after the proof.)

Given that $a_{11} \neq 0$, we can proceed as follows: for each $i > 1$, let $\lambda_i = a_{i1}/a_{11}$ and

subtract $\lambda_i \times$ (equation 1) from equation i.

This entire procedure of division, multiplication and subtraction is known as *one Elimination Step*. Since

$$a_{i1} - \lambda_i a_{11} = 0 \qquad \text{for } i > 1 \tag{1}$$

the elimination step leaves us with a system in which x_1 has zero coefficient in all equations other than the first. The first equation is unchanged. We have the system

$$a_{11}x_1 + a_{12}x_2 + \cdots + a_{1n}x_n = b_1 \tag{2}$$

$$\left.\begin{aligned} e_{22}x_2 + \cdots + e_{2n}x_n &= f_2 \\ \cdots \\ e_{n2}x_2 + \cdots + e_{nn}x_n &= f_n \end{aligned}\right\} \tag{3}$$

where $e_{ij} = a_{ij} - \lambda_i a_{1j}$ and $f_i = b_i - \lambda_i b_1$.

Now (3) is an equation system of order $n - 1$ and we shall prove in the next paragraph that its coefficient matrix (e_{ij}) is *nonsingular*. Given this, and the induction hypothesis, the values of $x_2, \ldots, x_n$ are uniquely determined. But then x_1 is uniquely determined by one step of back-substitution on (2).

It remains to show that (3) is a nonsingular system. Let $y_2, \ldots, y_n$ be scalars such that

$$e_{i2}y_2 + \cdots + e_{in}y_n = 0 \text{ for all } i > 1. \tag{4}$$

We wish to prove that $y_2 = \cdots = y_n = 0$. Now

$$e_{ij} = a_{ij} - \lambda_i a_{1j} = a_{ij} - a_{i1}a_{11}^{-1}a_{1j}.$$

Defining the scalar y_1 by

$$y_1 = -(a_{12}y_2 + \cdots + a_{1n}y_n)/a_{11},$$

we may rewrite (4) as

$$a_{i1}y_1 + a_{i2}y_2 + \cdots + a_{in}y_n = 0 \quad \text{for all } i > 1. \tag{5}$$

But by definition of y_1,

$$a_{11}y_1 + a_{12}y_2 + \cdots + a_{1n}y_n = 0. \tag{6}$$

Letting $\mathbf{y} = (y_1, y_2, \ldots, y_n)^{\mathrm{T}}$ we infer from (5) and (6) that $\mathbf{Ay} = \mathbf{0}$. Since $\mathbf{A}$ is nonsingular, $\mathbf{y} = \mathbf{0}$. QED

This proof may not look like a recipe for solving nonsingular systems, but it can be turned into one. There are three difficulties to surmount. First, the proof uses the somewhat indirect method of mathematical induction. This turns out to be no problem at all, thanks to back-substitution: more on this presently. Second, the very statement of the theorem begins, 'let **A** be a nonsingular matrix': how do we know before we start that we do not have a singular system? The answer is that we do not, and the solution procedure also acts as a test for singularity: but how this is done will not be revealed until the end of the next section, and we merely assure the reader that the systems of Example 3 and Exercises 2 and 3 are nonsingular.

The third difficulty lies in the 'without loss of generality' bit at the beginning of the proof. Given a nonsingular system, we cannot be sure that $a_{11} \neq 0$, but we can be sure that $a_{k1} \neq 0$ for some $k \geq 1$. Suppose that $a_{11} = a_{21} = 0$ but $a_{31} \neq 0$: we can then start the solution procedure by exchanging the first and third equations. We then have the same equations in a different order: the first row of the new coefficient matrix is the third row of **A**, and the top term on the right hand side is b_3. We call this step a *row interchange*. Since we have the same equations in a different order,

(*) a row interchange preserves nonsingularity.

We now proceed with an elimination step as in the proof of the theorem: for $i = 2, 3, \ldots$, we subtract λ_i times the (new) top equation from equation i, where λ_i is chosen so as to reduce the coefficient of x_1 to zero. This gives us a system like (2) and (3) above. The heavy algebra at the end of the proof was designed to show that

(**) an elimination step preserves nonsingularity.

We now get to work on the system (3), doing a row interchange if necessary and an elimination step. This leaves the first equation unchanged, and yields a system of order $(n - 2)$ to which the same procedure is applied. After $n - 1$ elimination steps, possibly interspersed with row interchanges, we obtain an upper triangular system. By (*), (**) and Theorem 1, this system is a NUT which we can solve by back-substitution. This method of obtaining a NUT is called *Gaussian elimination.*

In practice, one does not write down the equations each time. One omits the variables and operates on the rows of $n \times (n + 1)$ matrices, starting with the matrix $(\mathbf{A}, \mathbf{b})$. This is called the *augmented matrix* of the system.

Example 3 Solve the system of linear equations

$$
\begin{aligned}
2x_1 + 2x_2 + x_3 + 3x_4 &= 4 \\
6x_1 + 6x_2 + 5x_3 + 3x_4 &= 8 \\
2x_1 + x_2 + 3x_3 + 2x_4 &= 1 \\
4x_1 + x_2 + 2x_3 + 9x_4 &= 7.
\end{aligned}
$$

We start by writing down the augmented matrix

$$\left(\begin{array}{cccc|c} 2 & 2 & 1 & 3 & 4 \\ 6 & 6 & 5 & 3 & 8 \\ 2 & 1 & 3 & 2 & 1 \\ 4 & 1 & 2 & 9 & 7 \end{array}\right).$$

Since the first diagonal entry is not zero we may proceed to elimination without row interchange. We subtract multiples of the first row from the other rows to reduce entries in the first column to zero. This means subtracting 6/2 times the first row from the second, 2/2 times the first row from the third and 4/2 times the first row from the fourth. We obtain

$$\left(\begin{array}{cccc|c} 2 & 2 & 1 & 3 & 4 \\ 0 & 0 & 2 & -6 & -4 \\ 0 & -1 & 2 & -1 & -3 \\ 0 & -3 & 0 & 3 & -1 \end{array}\right)$$

Now the (2, 2) entry is zero and we need a row interchange. We exchange the second and third rows.

$$\left(\begin{array}{cccc|c} 2 & 2 & 1 & 3 & 4 \\ 0 & -1 & 2 & -1 & -3 \\ 0 & 0 & 2 & -6 & -4 \\ 0 & -3 & 0 & 3 & -1 \end{array}\right)$$

The next elimination subtracts three times the second row from the fourth: notice that nothing needs to be done to the third row since there is already a zero in the (3, 2) entry.

$$\left(\begin{array}{cccc|c} 2 & 2 & 1 & 3 & 4 \\ 0 & -1 & 2 & -1 & -3 \\ 0 & 0 & 2 & -6 & -4 \\ 0 & 0 & -6 & 6 & 8 \end{array}\right)$$

The final elimination gives us or our NUT:

$$\left(\begin{array}{cccc|c} 2 & 2 & 1 & 3 & 4 \\ 0 & -1 & 2 & -1 & -3 \\ 0 & 0 & 2 & -6 & -4 \\ 0 & 0 & 0 & -12 & -4 \end{array}\right).$$

By back-substitution,

$$\begin{aligned}
x_4 &= \tfrac{4}{12} = \tfrac{1}{3}, \\
x_3 &= (-4 + 6x_3)/2 = -1, \\
x_2 &= (-3 + x_4 - 2x_3)/(-1) = \tfrac{2}{3}, \\
x_1 &= (4 - 3x_4 - x_3 - 2x_2)/2 = 4/3.
\end{aligned}$$

Early in the proof of Theorem 1, and in the ensuing discussion, we gave the recipe for the elimination step in terms of operations on equations. What we actually do, as in Example 3, is to operate on matrices. In subsequent chapters we shall give recipes in algebraic form and it is worth the reader's while to get used to this now. Let (α_{ij}) denote the matrix which we write down immediately before the rth elimination step and (β_{ij}) the one immediately after it. Then

$$\begin{aligned}
\beta_{ij} &= \alpha_{ij} - \alpha_{ir}\alpha_{rr}^{-1}\alpha_{rj}, && \text{for } i > r \text{ and all } j, \\
\beta_{ij} &= \alpha_{ij}, && \text{for } i \le r \text{ and all } j.
\end{aligned}$$

There is no point in learning this formula by heart but the reader should not proceed until he has convinced himself that this is indeed the instruction for the rth elimination step.

Theorem 1.4.2 was the first of our two Laws of Linear Dependence. We can now state and prove the other one.

THEOREM 3 (Second Law of Linear Dependence)
Any $n + 1$ n-vectors are linearly dependent.

PROOF Choose $n + 1$ n-vectors: let the first n form the square matrix $\mathbf{A}$ and denote the other one by $\mathbf{b}$. We wish to find a non-zero $(n + 1)$-vector $(\mathbf{z}\colon \lambda)$ such that

$$\mathbf{Az} + \lambda\mathbf{b} = \mathbf{0}.$$

If $\mathbf{A}$ is singular there exists a non-zero n-vector $\mathbf{y}$ such that $\mathbf{Ay} = \mathbf{0}$. If $\mathbf{A}$ is nonsingular there exists, by Theorem 2, a unique n-vector $\mathbf{x}$ such that $\mathbf{Ax} = \mathbf{b}$. Set $\mathbf{z} = \mathbf{y}$ and $\lambda = 0$ in the first case, and $\mathbf{z} = \mathbf{x}$ and $\lambda = -1$ in the second. QED

Exercises

1. Solve the system

$$\begin{aligned}
2x_1 - 3x_2 + 6x_3 &= 8 \\
- x_2 - 7x_3 &= 1 \\
-2x_3 &= 6.
\end{aligned}$$

2. Solve the system

$$\begin{aligned} x_2 + x_3 + x_4 &= 2 \\ 2x_1 + 3x_2 - 3x_3 + x_4 &= 4 \\ x_1 - x_2 + 5x_3 - 3x_4 &= 3 \\ x_2 + 8x_3 &= 6. \end{aligned}$$

3. Solve the system

$$\begin{aligned} x_1 + 4x_2 + 2x_3 + 2x_4 + 3x_5 &= 5 \\ 2x_1 + 6x_2 + x_3 + x_4 + 4x_5 &= 3 \\ 3x_1 + 8x_2 + 7x_4 + 2x_5 &= 4 \\ x_1 + 3x_4 + x_5 &= 1 \\ 3x_1 + 9x_2 + x_3 + 5x_4 + 8x_5 &= 1. \end{aligned}$$

4. Prove that if **A** and **B** are upper triangular $n \times n$ matrices, so is **AB**. Hence or otherwise prove that this result holds when 'upper' is replaced by 'lower'.

2.2 The inverse matrix

The major question arising from the last section is: how can we tell if a given system of equations is singular? We shall answer this at the end of the section. Meanwhile, it is helpful to gather as much useful information as we can about nonsingular systems. The main theorem is as follows.

THEOREM 1 Given a square matrix **A**, the following three statements are equivalent:
(α) **A** is nonsingular;
(β) for any **b**, the system $\mathbf{Ax} = \mathbf{b}$ has a solution;
(γ) for any **b**, the system $\mathbf{Ax} = \mathbf{b}$ has a unique solution.

Let us see how much of this has already been established. (α) implies (γ) by Theorem 2.1.2, (γ) implies (β) trivially and we can obtain (α) from (γ) by setting $\mathbf{b} = \mathbf{0}$. Thus (α) and (γ) are equivalent statements and imply (β). What we have not yet proved is that (β) is a *sufficient* condition for nonsingularity.

To tackle this, we introduce the notion of an *inverse matrix* and prove two theorems about inverses. We then return to the proof of Theorem 1.

Let $\mathbf{A}$ be a nonsingular matrix of order n and let $\mathbf{u}_1, \ldots, \mathbf{u}_n$ denote the columns of $\mathbf{I}_n$. By (γ) there exists for each j exactly one n-vector $\mathbf{v}_j$ such that $\mathbf{A}\mathbf{v}_j = \mathbf{u}_j$. Hence there exists exactly one matrix $\mathbf{V}$ such that $\mathbf{AV} = \mathbf{I}$. This $\mathbf{V}$ is called the *inverse* of $\mathbf{A}$ and is denoted by $\mathbf{A}^{-1}$. Thus

$$\mathbf{A}\mathbf{A}^{-1} = \mathbf{I}.$$

Since matrix multiplication is not in general commutative, we have not yet proved that

$$\mathbf{A}^{-1}\mathbf{A} = \mathbf{I}.$$

The statement is, however, correct. The following theorem says this, and more.

THEOREM 2 Let $\mathbf{A}$ be a square matrix. There exists a matrix $\mathbf{B}$ such that $\mathbf{BA} = \mathbf{I}$ if and only if $\mathbf{A}$ is nonsingular, in which case the only such $\mathbf{B}$ is $\mathbf{A}^{-1}$.

PROOF Let $\mathbf{BA} = \mathbf{I}$. If $\mathbf{Ax} = \mathbf{0}$ then $\mathbf{x} = \mathbf{BAx} = \mathbf{0}$. Thus $\mathbf{A}$ is nonsingular and

$$\mathbf{B} = \mathbf{B}(\mathbf{A}\mathbf{A}^{-1}) = (\mathbf{BA})\mathbf{A}^{-1} = \mathbf{A}^{-1}.$$

This proves 'only if' and 'in which case'.

Conversely, let $\mathbf{A}$ be nonsingular. We want to prove that $\mathbf{A}^{-1}\mathbf{A} = \mathbf{I}$. Let $\mathbf{x}$ be any vector and let $\mathbf{y} = (\mathbf{A}^{-1}\mathbf{A} - \mathbf{I})\mathbf{x}$. By Theorem 1.4.1 it suffices to show that $\mathbf{y} = \mathbf{0}$. Now

$$\mathbf{Ay} = (\mathbf{A}\mathbf{A}^{-1})(\mathbf{Ax}) - \mathbf{Ax} = \mathbf{IAx} - \mathbf{Ax} = \mathbf{0}$$

so $\mathbf{y} = \mathbf{0}$ by nonsingularity of $\mathbf{A}$. QED

Theorem 2 is known as the Left Inverse Property of nonsingular matrices. The following theorem is the Right Inverse Property.

THEOREM 3 Let $\mathbf{A}$ be a square matrix. There exists a matrix $\mathbf{B}$ such that $\mathbf{AB} = \mathbf{I}$ if and only if $\mathbf{A}$ is nonsingular, in which case the only such $\mathbf{B}$ is $\mathbf{A}^{-1}$.

PROOF We proved the 'if' and 'in which case' parts of this theorem in the process of defining the inverse. It remains to prove 'only if'.

Let $\mathbf{AB} = \mathbf{I}$. Then we can apply Theorem 2 *with the roles of* $\mathbf{A}$ *and* $\mathbf{B}$ *reversed*, inferring that $\mathbf{B}$ is nonsingular with inverse $\mathbf{A}$. But then $\mathbf{BA} = \mathbf{I}$, so $\mathbf{A}$ is nonsingular by *direct* application of Theorem 2.

QED

The reversion-of-rôles argument establishes another important fact: if $\mathbf{A}$ is nonsingular so is $\mathbf{A}^{-1}$ and

$$(\mathbf{A}^{-1})^{-1} = \mathbf{A}.$$

The 'only ifs' in Theorems 2 and 3 are all-important. They demonstrate that it is not possible to speak of the inverse of a singular matrix.

We can now complete the proof of Theorem 1. Suppose that (β) is true. Then for each column $\mathbf{u}_j$ of $\mathbf{I}$, there exists a vector $\mathbf{v}_j$ such that $\mathbf{A}\mathbf{v}_j = \mathbf{u}_j$. Hence there exists a matrix $\mathbf{V}$ such that $\mathbf{AV} = \mathbf{I}$, and $\mathbf{A}$ is nonsingular by Theorem 3.

The next two theorems show how inversion interacts with transposition and multiplication.

THEOREM 4 If $\mathbf{A}$ is nonsingular so is $\mathbf{A}^{\mathrm{T}}$, and $(\mathbf{A}^{\mathrm{T}})^{-1} = (\mathbf{A}^{-1})^{\mathrm{T}}$.

PROOF Transposing $\mathbf{I} = \mathbf{A}\mathbf{A}^{-1}$ we have $\mathbf{I} = (\mathbf{A}^{-1})^{\mathrm{T}}\mathbf{A}^{\mathrm{T}}$. Now apply Theorem 2 with $\mathbf{A}$ replaced by $\mathbf{A}^{\mathrm{T}}$. QED

THEOREM 5 If $\mathbf{A}$ and $\mathbf{B}$ are nonsingular matrices of the same order, so is $\mathbf{AB}$, and

$$(\mathbf{AB})^{-1} = \mathbf{B}^{-1}\mathbf{A}^{-1}. \tag{1}$$

PROOF $\mathbf{ABB}^{-1}\mathbf{A}^{-1} = \mathbf{AIA}^{-1} = \mathbf{AA}^{-1} = \mathbf{I}.$ QED

Notice the reversion of order in (1), similar to that in transposition of a product. Notice also that (1) may be iterated:

$$(\mathbf{ABCDE})^{-1} = \mathbf{E}^{-1}\mathbf{D}^{-1}\mathbf{C}^{-1}\mathbf{B}^{-1}\mathbf{A}^{-1}.$$

Notice further that while we now have an algebra of inversion, we have neither an arithmetic nor a motivation. We deal with the latter point first.

Let $\mathbf{Ax} = \mathbf{b}$, with $\mathbf{A}$ nonsingular. Premultiplying by $\mathbf{A}^{-1}$ we have $\mathbf{x} = \mathbf{A}^{-1}\mathbf{b}$. Thus one way to solve the system $\mathbf{Ax} = \mathbf{b}$ is to compute $\mathbf{A}^{-1}$ and postmultiply it by $\mathbf{b}$. *This is less efficient than Gaussian elimination.*

One reason why one might want to compute $\mathbf{A}^{-1}$ is to solve a sequence of linear equation systems, all with coefficient matrix $\mathbf{A}$ but with different right-hand sides. Thus $\mathbf{A}^{-1}$ is storable information which can be used to calculate $\mathbf{A}^{-1}\mathbf{b}$ for any given $\mathbf{b}$. (In fact one can instruct a computer to store equivalent information in a much more efficient way, but we shall not go into that here.) Also, by Theorem 4, we can use the information $\mathbf{A}^{-1}$ to solve any system whose coefficient matrix is $\mathbf{A}^{\mathrm{T}}$: if $\mathbf{A}^{\mathrm{T}}\mathbf{p} = \mathbf{c}$ then $\mathbf{p} = (\mathbf{A}^{-1})^{\mathrm{T}}\mathbf{c}$.

We have not yet explained *how* to invert a matrix, but one popular method is implicit in the definition. For each j, let $\mathbf{u}_j$ denote column j of the identity matrix and solve the system $\mathbf{Ax} = \mathbf{u}_j$ by elimination and substitution: the unique solution is column j of $\mathbf{A}^{-1}$. When we invert matrices in Chapter 3, we shall not use that method but a slightly different technique called pivoting. But we have already said enough to justify our earlier statement that inversion is an inefficient way of solving just one system: inverting $\mathbf{A}$ requires the solution of the n systems $\mathbf{Ax} = \mathbf{u}_j$ $(j = 1, \ldots, n)$.

In the case $n = 2$, there is a formula for the inverse which is simple enough to remember.

THEOREM 6 Let $\mathbf{A}$ be a 2×2 matrix and let

$$\delta = a_{11}a_{22} - a_{12}a_{21}.$$

Then $\mathbf{A}$ is nonsingular if and only if $\delta \neq 0$, in which case

$$\mathbf{A}^{-1} = (1/\delta)\begin{pmatrix} a_{22} & -a_{12} \\ -a_{21} & a_{11} \end{pmatrix}. \tag{2}$$

PROOF If $\delta \neq 0$ we can write the right-hand side of (2) and premultiply it by $\mathbf{A}$ to obtain $\mathbf{I}_2$. This proves 'if' and 'in which case'. To prove 'only if' let $\mathbf{A}$ be nonsingular. Then, for any $\mathbf{b}$, the system $\mathbf{Ax} = \mathbf{b}$ can be solved by Gaussian elimination. We know that the (2, 2) entry after the elimination step is not zero: call it γ. If $a_{11} \neq 0$, then

$$\gamma = a_{22} - a_{21}a_{11}^{-1}a_{12} = \delta/a_{11}.$$

If $a_{11} = 0$ (the row interchange case) then $\gamma = -\delta/a_{21}$. In each case $\delta \neq 0$. QED

Example 1 If $\mathbf{A} = \begin{pmatrix} 3 & 2 \\ 5 & 6 \end{pmatrix}$ then $\delta = 3 \times 6 - 2 \times 5 = 8$, so

$$\mathbf{A}^{-1} = (\tfrac{1}{8})\begin{pmatrix} 6 & -2 \\ -5 & 3 \end{pmatrix} = \begin{pmatrix} \frac{3}{4} & -\frac{1}{4} \\ -\frac{5}{8} & \frac{3}{8} \end{pmatrix}.$$

Although inversion of matrices cannot be done by Gaussian elimination on one square system, the theory of this section can tell us something about that process.

The rth elimination step is

$$\beta_{ij} = \alpha_{ij} - \alpha_{ir}\alpha_{rr}^{-1}\alpha_{rj}, \quad \text{for } i > r \text{ and all } j,$$
$$\beta_{ij} = \alpha_{ij}, \quad \text{for } i \le r \text{ and all } j.$$

Now this is equivalent to premultiplying (α_{ij}) by a matrix $\mathbf{E}_r$ which differs from $\mathbf{I}$ only in the rth *column*. Specifically, column r of $\mathbf{E}_r$ is

$$(0 \quad 0 \quad \cdots \quad 0 \quad 1 \quad -\lambda_{r+1} \quad \cdots \quad -\lambda_n)^{\mathrm{T}},$$

where the 1 is the rth component and $\lambda_i = \alpha_{ir}/\alpha_{rr}$ for $i > r$. Verification that premultiplying by $\mathbf{E}_r$ does the rth elimination step is a simple matter of applying the rule

$$\text{row } i \text{ of } (\mathbf{AB}) = (\text{row } i \text{ of } \mathbf{A})\mathbf{B}.$$

The reader should check this both in general and with numerical examples.

The main theoretical point here is that $\mathbf{E}_r$ is a lower triangular matrix with 1's down the diagonal and is therefore nonsingular (Theorem 2.1.1). Similarly, interchanging rows r and s is equivalent to premultiplication by the nonsingular matrix $\mathbf{J}$ obtained by interchanging rows r and s of $\mathbf{I}$.

To illustrate this, let us write down in matrix algebra the story of an elimination process. Suppose we solve a nonsingular 3×3 system $\mathbf{Ax} = \mathbf{b}$. This takes two elimination steps: let us suppose there is one row interchange, occurring between these two steps. The matrices we write, starting with the augmented matrix $\mathbf{C} = (\mathbf{A}, \mathbf{b})$ are

$$\mathbf{C},\ \mathbf{E}_1\mathbf{C},\ \mathbf{JE}_1\mathbf{C},\ \mathbf{E}_2\,\mathbf{JE}_1\mathbf{C}$$

where $\mathbf{E}_1$, $\mathbf{E}_2$ and $\mathbf{J}$ (the row-interchange premultiplier) are nonsingular matrices. We may write this sequence

$$\mathbf{C},\ \mathbf{G}_1\mathbf{C},\ \mathbf{G}_2\mathbf{C},\ \mathbf{G}_3\mathbf{C}$$

where $\mathbf{G}_1 = \mathbf{E}_1$, $\mathbf{G}_2 = \mathbf{JE}_1$, $\mathbf{G}_3 = \mathbf{E}_2\,\mathbf{JE}_1$: notice the order of multiplication. The matrices $\mathbf{G}_1$, $\mathbf{G}_2$, $\mathbf{G}_3$ are called *row operators* and are *nonsingular* by Theorem 5.

What has this to do with inversion? Notice that $\mathbf{G}_3$ is *not* in general $\mathbf{A}^{-1}$: if it were, we could read off the solution $\mathbf{x}$ without back-substitution, and we know that this is not in general the case. What *is* relevant is that the original augmented matrix $\mathbf{C}$ can in principle be recovered from $\mathbf{G}_3\mathbf{C}$ by multiplication by $\mathbf{G}_3^{-1}$. This means that the equation systems we write (implicitly) in the process of Gaussian elimination are *equivalent systems*.

We showed in the last section that *if* an $n \times n$ system is nonsingular, *then* it can be reduced to a NUT. But since each row operator $\mathbf{G}$ in Gaussian elimination can be inverted, the converse is also true: *if* $\mathbf{GA} = \mathbf{U}$ where $\mathbf{U}$ is nonsingular, *then* the matrix $\mathbf{A} = \mathbf{G}^{-1}\mathbf{U}$ is nonsingular. Thus *singular* systems are precisely those for which Gaussian elimination *fails* to produce a NUT. There are two possible sources of failure: either the illegality of dividing by zero prevents us from performing all $n - 1$ elimination steps, or all steps can be performed and the (n, n) entry of the resulting upper triangular matrix is zero. The following singularity test covers both cases:

> an $n \times n$ matrix is singular if and only if there is some $k \leq n$ such that, after $k - 1$ elimination steps, we have $\alpha_{ik} = 0$ for all $i \geq k$.

Example 2

$$\text{Let } \mathbf{A} = \begin{pmatrix} 1 & 2 & 5 & 1 \\ 2 & 3 & 6 & 1 \\ 0 & 1 & 4 & 2 \\ 5 & 6 & 1 & 8 \end{pmatrix}.$$

After two elimination steps we have

$$\begin{pmatrix} 1 & 2 & 5 & 1 \\ 0 & -1 & -4 & -1 \\ 0 & 0 & 0 & 1 \\ 0 & 0 & 0 & 9 \end{pmatrix}$$

Here $\alpha_{33} = \alpha_{43} = 0$ so $\mathbf{A}$ is singular.

We have now shown to *detect* a singular system. To *solve* such systems we need the theory and methods of the next two sections.

Exercises

1. Invert the matrices

$$\begin{pmatrix} 8 & -2 \\ -1 & 4 \end{pmatrix} \text{ and } \begin{pmatrix} 3 & 5 \\ 2 & 1 \end{pmatrix}.$$

2. Show that the following system is singular.

$$\begin{aligned} x_1 + x_2 + 3x_3 &= 4 \\ 2x_1 - x_2 + 2x_3 &= 3 \\ 7x_1 - 8x_2 + x_3 &= 3 \end{aligned}$$

3. Show that in cases where Gaussian elimination can be performed without row interchanges, the row operators are lower triangular.

2.3 Dimension

We begin with a simple logical point about the main theorem of the preceding section, Theorem 2.2.1. This theorem is just as much a statement about singular matrices as about nonsingular ones.

Let $\mathbf{A}$ be a singular $n \times n$ matrix. By definition, there exists some non-zero n-vector $\mathbf{z}$ such that $\mathbf{Az} = 0$. But then $\mathbf{A}(2\mathbf{z}) = \mathbf{0}$, $\mathbf{A}(3\mathbf{z}) = \mathbf{0}$ and so on, so that the set $\{\mathbf{x} \in R^n \colon \mathbf{Ax} = \mathbf{0}\}$ is infinite. We also know (Theorem 2.2.1: (β) implies (α)) that there must be some $\mathbf{b}$ for which the set $\{\mathbf{x} \in R^n \colon \mathbf{Ax} = \mathbf{b}\}$ is empty.

This suggests that in a singular system, almost anything can happen. In fact, the one thing that *cannot* happen is what *always* happens in nonsingular systems, namely a unique solution. For

$$\text{if } \mathbf{Ax} = \mathbf{b} \text{ and } \mathbf{Az} = \mathbf{0} \text{ then } \mathbf{A}(\mathbf{x} + \mathbf{z}) = \mathbf{b}.$$

So solving n equations in n unknowns without imposing the assumption of nonsingularity is a complicated business. In fact, it is no more and no less complicated than solving m equations in n unknowns, where m and n may be different. So we proceed to the $m \times n$ case, via the geometrical notion of a linear subspace. What is geometrical about it will be explained at the end of the section. Its relevance to equation systems will become clear much sooner.

We say that a non-empty set L of n-vectors is a *linear subspace* of R^n if, given $\mathbf{x}$ and $\mathbf{y}$ in L and scalars λ and μ, we have $\lambda\mathbf{x} + \mu\mathbf{y} \in L$. Notice that this implies that *any* linear combination of vectors in L is also in L: for

$$\lambda\mathbf{x} + \mu\mathbf{y} + \nu\mathbf{z} = (\lambda\mathbf{z} + \mu\mathbf{y}) + \nu\mathbf{z}$$

and so on.

Two examples of linear subspaces are immediate: the singleton $\{\mathbf{0}\}$ and R^n itself. To get some more interesting ones, we introduce the 'range' and 'kernel' of a matrix. We define the *range* of a matrix $\mathbf{A}$ to be the set of all linear combinations of the columns of $\mathbf{A}$, and denote it by rng $\mathbf{A}$. We define the *kernel* of $\mathbf{A}$ to be the set of all vectors $\mathbf{z}$ for which $\mathbf{Az} = \mathbf{0}$ and denote it by ker $\mathbf{A}$.

THEOREM 1 Given an $m \times n$ matrix $\mathbf{A}$:
(i) rng $\mathbf{A}$ is a linear subspace of R^m;
(ii) ker $\mathbf{A}$ is a linear subspace of R^n.
PROOF Let $\mathbf{x}$ and $\mathbf{y}$ be n-vectors and let λ and μ be scalars. Then

$$\lambda(\mathbf{Ax}) + \mu(\mathbf{Ay}) = \mathbf{A}(\lambda\mathbf{x} + \mu\mathbf{y})$$

which proves (i). If $\mathbf{Ax} = \mathbf{Ay} = \mathbf{0}$, then

$$\mathbf{A}(\lambda\mathbf{x} + \mu\mathbf{y}) = \mathbf{0}$$

which proves (ii). QED

Example 1 Let $L = \{\mathbf{x} \in R^4\colon x_1 + x_2 = x_3 + x_4 = 0\}$.
Then L is the kernel of the 2×4 matrix

$$\begin{pmatrix} 1 & 1 & 0 & 0 \\ 0 & 0 & 1 & 1 \end{pmatrix}.$$

Also notice that the 4-vector $\mathbf{x}$ belongs to L if and only if it is of the form

$$(\alpha, -\alpha, \beta, -\beta)^{\mathrm{T}}$$

for some scalars α and β. Thus L is the range of the 4×2 matrix

$$\begin{pmatrix} 1 & 0 \\ -1 & 0 \\ 0 & 1 \\ 0 & -1 \end{pmatrix}.$$

We have expressed the linear subspace L as the range of a matrix and as the kernel of another matrix. This is a general property of linear subspaces as we shall show in Section 4.1.

Let $\mathbf{A}$ be an $m \times n$ matrix, $\mathbf{b}$ an m-vector. The system

$$\mathbf{Ax} = \mathbf{b}$$

has a solution if and only if $\mathbf{b}$ is in the range of $\mathbf{A}$. If a solution $\mathbf{x}^0$ exists, the set of all solutions consists of all vectors of the form $\mathbf{x}^0 + \mathbf{z}$, where $\mathbf{z}$ is an arbitrary member of the kernel of $\mathbf{A}$. These propositions follow immediately from the definitions of range and kernel and suggest that we should get our facts right about ranges and kernels before tackling $m \times n$ systems. And the 'general property' mentioned in Example 2 suggest that we may as well study linear subspaces in general. That is the subject of this section. Our main tools will be the Laws of Linear Dependence.

One question we can ask about a linear subspace L of R^n is: how big is it? One answer is immediate: L is either $\{\mathbf{0}\}$ or an infinite set. For if $\mathbf{z}$ is in L so are $2\mathbf{z}$, $3\mathbf{z}$ and so on. But there is another notion of size which is more useful for our purposes.

First consider the case where L is not $\{\mathbf{0}\}$. Then some non-zero vector, say $\mathbf{z}$, is in L. If $\lambda\mathbf{z} = \mathbf{0}$ then $\lambda = 0$. On the other hand, the Second Law tells us that we cannot choose a set of $n + 1$ linearly independent vectors in L. Choose a positive integer k and ask the question, can k linearly independent n-vectors be selected from L? The largest k for which the answer is 'Yes' is called the *dimension* of L; it is positive and cannot exceed n. It remains to consider the case where $L = \{\mathbf{0}\}$: in this case we define the dimension to be zero.

We can summarise all this by saying that the dimension of a linear subspace L is the *maximal* number of linearly independent members of L: it is denoted by $\dim L$. The ratiocinations of the preceding paragraph were intended to do three things. First, to show that $\dim L \geq 0$ with equality if and only if $L = \{\mathbf{0}\}$. Second, we showed that the dimension of a linear subspace of R^n cannot exceed n. Third, we attempted to give the reader some idea of what the term 'maximal' means: from now on we shall use it without fuss, but the reader should remember that it is the Second Law which keeps our 'maximal numbers' finite.

The next theorem shows that dimension is a sensible notion of size.

THEOREM 2 Let K and L be linear subspaces of R^n such that $K \subset L$. Then $\dim K \leq \dim L$, with equality if and only if $K = L$.

PROOF Let $\dim K = k$. If $K = L$, $\dim L = k$. Now let K be a proper subset of L. We want to show that $\dim L > k$.

If $k = 0$ the result is obvious. Now let $k > 0$ and let $\mathbf{x}_1, \ldots, \mathbf{x}_k$ be k linearly independent vectors in K. Let $\mathbf{x}_0$ be a member of L which is not in K. Since K is a linear subspace, $\mathbf{x}_0$ is not a linear combination of $\mathbf{x}_1, \ldots, \mathbf{x}_k$. By the First Law of Linear Dependence, $\mathbf{x}_0, \mathbf{x}_1, \ldots, \mathbf{x}_k$ are $1 + k$ linearly independent members of L. QED

By the Second Law, and the fact that nonsingular matrices exist, the dimension of R^n is n. Theorem 2, with $L = R^n$, shows that the *only* n-dimensional linear subspace of R^n is R^n itself.

Example 2 Theorem 2 says that if K is a proper subset of L then dim $K <$ dim L. The converse is false. In R^3 let K be the linear subspace of all vectors with zero first and second components, L the linear subspace of all vectors with zero third component. Then L is simply R^2 with zeroes attached, so dim $L = 2$. Similarly dim $K = 1$. But K is not contained in L.

Let L be a linear subspace of R^n and let $\mathbf{y}_1, \ldots, \mathbf{y}_m$ be members of L. These vectors are said to *span* L if any member of L can be expressed as a linear combination of them: this is just the same as saying that L is the range of the matrix with columns $\mathbf{y}_1, \ldots, \mathbf{y}_m$. The next two theorems relate dimension to spanning.

THEOREM 3 Let L be a linear subspace of R^n of dimension $d > 0$. Let $\mathbf{y}_1, \ldots, \mathbf{y}_m$ span L. Then $m \geq d$, with equality if and only if $\mathbf{y}_1, \ldots, \mathbf{y}_m$ are linearly independent.

PROOF By definition of dimension we can choose an $n \times d$ matrix $\mathbf{X}$ whose columns are in L and are linearly independent. Let $\mathbf{y}_1, \ldots, \mathbf{y}_m$ form the $n \times m$ matrix $\mathbf{Y}$: then $L = \text{rng } \mathbf{Y}$. Since each column of $\mathbf{X}$ is in L there exists an $m \times d$ matrix $\mathbf{Z}$ such that $\mathbf{X} = \mathbf{YZ}$. If $\mathbf{Zp} = \mathbf{0}$ then $\mathbf{Xp} = \mathbf{0}$ so $\mathbf{p} = \mathbf{0}$. Thus the columns of $\mathbf{Z}$ are linearly independent, and $m \geq d$ by the Second Law.

If the columns of $\mathbf{Y}$ are linearly independent then $m \leq d$ by definition of dimension: thus $m = d$, by the preceding paragraph. If the columns of $\mathbf{Y}$ are linearly dependent then one of them can be expressed as a linear combination of the others. Since a linear combination of linear combinations is a linear combination (the associative law of matrix multiplication) we can then express any member of L as a linear combination of $m - 1$ columns of $\mathbf{Y}$. Then the $m \geq d$ argument goes through with m replaced by $m - 1$, so $m \geq d + 1$.

QED

THEOREM 4 Let L, d be as in Theorem 3 and let $\mathbf{x}_1, \ldots, \mathbf{x}_k$ be k linearly independent vectors in L. Then $k \leq d$, with equality if and only if $\mathbf{x}_1, \ldots, \mathbf{x}_k$ span L.

PROOF Of course $k \leq d$ by definition of dimension, but we must also prove the 'equality if and only if'. We start from scratch. Let $\mathbf{X}$ be the matrix formed by $\mathbf{x}_1, \ldots, \mathbf{x}_k$ and let its range be K. Then K is spanned by k linearly independent vectors, so its dimension is k by Theorem 3. But $K \subset L$, with equality only if $L = \text{rng } \mathbf{X}$. The required result now follows from Theorem 2. QED

We are now in a position to talk about bases, which are as fundamental as their name suggests. Let L be a linear subspace of R^n of dimension $d > 0$. Let $\mathbf{B}$ be a matrix with n rows. We say that $\mathbf{B}$ is a *basis* for L if *all three* of the following statements are true.

(i) The columns of $\mathbf{B}$ belong to L and are linearly independent.
(ii) The columns of $\mathbf{B}$ span L.
(iii) $\mathbf{B}$ has d columns.

By Theorems 3 and 4, *any two of these statements imply the third.* Thus $\mathbf{B}$ is a basis for L if any two of (i), (ii) and (iii) are true.

The first point to notice about this development is that it enables us to recognise dimensions when we see them. Suppose we have a linear subspace L of R^4, and we know that it is spanned by two linearly independent 4-vectors. Since (i) and (ii) imply (iii), $\dim L = 2$. This is precisely what happened in Example 1.

The second point is that any linear subspace of R^n, other than $\{\mathbf{0}\}$, has a basis. For if $d > 0$, the existence of a matrix satisfying (i) and (iii) is guaranteed by definition of dimension.

Now consider the case where $L = R^n$. By definition of basis, and the fact that $\dim R^n = n$, a matrix $\mathbf{A}$ with n rows is a basis for R^n if any two (and hence all) of the following hold.

(i^0) The columns of $\mathbf{A}$ are linearly independent.
(ii^0) The columns of $\mathbf{A}$ span R^n.
(iii^0) $\mathbf{A}$ is square.

By (i^0) and (iii^0), a basis for R^n is the same thing as an $n \times n$ nonsingular matrix. By (ii^0) and (iii^0), a basis for R^n is the same thing as a square matrix whose range is R^n. Thus a square matrix of order n is nonsingular if and only if its range is R^n.

We know this already: it is the equivalence of (α) and (β) in Theorem 2.2.1. So we now have two proofs of this. But they are not really very different. The earlier one used inversion, and the later the Laws of Linear Dependence, but both stem from Gaussian elimination, Theorem 2.1.2.

We may summarise the properties of bases, and Theorems 3 and 4, as follows.

(I) The only linear subspace of R^n that does not have a basis is $\{\mathbf{0}\}$.
(II) Let L be a linear subspace of R^n which is neither $\{\mathbf{0}\}$ nor R^n. A basis for L is a matrix whose columns span L and are linearly independent. Any two such matrices have the same number of columns. This number is the dimension of L: call it d. Then $0 < d < n$. Any matrix whose columns span L has *at least* d columns. Any matrix whose columns are in L and are linearly independent has *at most* d columns.

(III) A basis for R^n is a nonsingular $n \times n$ matrix.

Two points about the literature should be made here. First, it is more common to define a basis as a set rather than a matrix. But the definition used here is convenient for our purposes and we shall use it consistently. Second, some books regard the empty set as a linear subspace and define dim $\varnothing$ to be -1. In this book, linear subspaces are defined to be non-empty.

We now give another definition involving maximality. The *rank* of a matrix is its maximal number of linearly independent columns. The last theorem of this section will explain the relation between rank and basis.

There are various important weak inequalities ($\leq$ and $\geq$) connected with rank. We state the main ones at the end of the next section, but we can state three of them now.

Let $\mathbf{A}$ be an $m \times n$ matrix, and let its rank be r. By definition of rank, $r \geq 0$ and $r \leq n$: notice that $r = 0$ if and only if $\mathbf{A}$ is a zero matrix, while $r = n$ if and only if the columns of $\mathbf{A}$ are linearly independent. The third inequality comes from the Second Law of Linear Dependence: $r \leq m$. Thus r is a non-negative integer which cannot exceed the lesser of m and n. In the sequel, we shall denote the rank of $\mathbf{A}$ by rk $\mathbf{A}$; our inequalities may be written

$$0 \leq \text{rk } \mathbf{A} \leq \min (m, n).$$

THEOREM 5 Let $\mathbf{A}$ be a matrix of rank $r > 0$. Let the matrix $\mathbf{B}$ be formed by r linearly independent columns of $\mathbf{A}$. Then $\mathbf{B}$ is a basis for rng $\mathbf{A}$.

PROOF By maximality of r and the First Law of Linear Dependence, any column of $\mathbf{A}$ is a linear combination of the columns of $\mathbf{B}$. Hence, by the associative law of matrix multiplication, any member of rng $\mathbf{A}$ is a linear combination of the columns of $\mathbf{B}$. Thus $\mathbf{B}$ is a basis for rng $\mathbf{A}$. QED

It follows from Theorem 5 that if rk $\mathbf{A} > 0$ then rk $\mathbf{A}$ is the dimension of rng $\mathbf{A}$. But if rk $\mathbf{A} = 0$ then $\mathbf{A}$ is a zero matrix and its range has dimension zero. Hence

the rank of a matrix is the dimension of its range.

We come at last to the geometry of linear subspaces. Any linear subspace of R^2 has dimension 0, 1 or 2. Now a 1-dimensional subspace of R^2 (and indeed of any R^n) is the set of all scalar multiples of a single non-zero vector. Thus the linear subspaces of R^2 are $\{\mathbf{0}\}$, R^2 and the lines through the origin. The linear subspaces of R^3 are $\{\mathbf{0}\}$, R^3 and lines and planes through the origin.

Exercises

1. Show that if L and M are linear subspaces of R^n then $L \cap M$ is a linear subspace, but $L \cup M$ is not in general a linear subspace.

2. Let S be a (possibly infinite) set in R^n which owns at least one non-zero vector. We say that an n-vector x is a linear combination of the members of S if it is a linear combination of the members of some finite subset of S. Let $L(S)$ denote the set of all linear combinations of members of S. Prove
   (i) $L(S)$ is a linear subspace;
   (ii) if M is any linear subspace which contains S, then $L(S) \subset M$;
   (iii) $L(S)$ has a basis whose columns belong to S.
   The set S is known as the linear subspace generated by S. Illustrate (i), (ii) and (iii) in the case where

$$S = \{\mathbf{x} \in R^3 \colon x_1 > 2,\ x_2 = -1 \text{ and } x_3 = 0\}.$$

3. Given linear subspaces L and M of R^n, we define their *sum* $L + M$ to be the set of all vectors of the form $\mathbf{x} + \mathbf{y}$, where $\mathbf{x} \in L$ and $\mathbf{y} \in M$.
   (i) If $L = \text{rng } \mathbf{A}$ and $M = \text{rng } \mathbf{B}$, find a matrix whose range is $L + M$.
   (ii) Show that $\dim (L + M) \le \dim L + \dim M$.
   (iii) Show that $L + M$ is the linear subspace generated by $L \cup M$.

4. Let $\mathbf{B}$ be a basis for a linear subspace L of R^n. Show that the matrix $\mathbf{C}$ is a basis for L if and only if $\mathbf{C} = \mathbf{BA}$ for some nonsingular $\mathbf{A}$.

2.4 General systems

We now pull together the threads of the last two sections. Given an $m \times n$ matrix $\mathbf{A}$ we can describe *four* linear subspaces associated with $\mathbf{A}$. The first two, introduced in the preceding section, are the range and kernel of $\mathbf{A}$. The others are the range and kernel of $\mathbf{A}^T$. Notice that rng $\mathbf{A}$ and ker $\mathbf{A}^T$ are linear subspaces of R^m, while rng $\mathbf{A}^T$ and ker $\mathbf{A}$ are linear subspaces of R^n. It turns out the dimensions of these four subspaces are related in a very simple way, stated in Theorem 1. The important existence theorems on linear equation systems follow from Theorem 1: these are Theorems 3 and 4. The solution procedure, illustrated in the examples, is suggested by the proof of Theorem 1. We end with some more inequalities concerning rank.

We already have names for two of the dimensions we are interested in: the dimension of rng $\mathbf{A}$ is rk $\mathbf{A}$, and the dimension of rng $\mathbf{A}^{\mathrm{T}}$ is of course rk $\mathbf{A}^{\mathrm{T}}$. The dimension of the kernel of $\mathbf{A}$ is called the *nullity* of $\mathbf{A}$ and is denoted by nul $\mathbf{A}$. Similarly,

$$\text{nul } \mathbf{A}^{\mathrm{T}} = \dim \ker \mathbf{A}^{\mathrm{T}}.$$

THEOREM 1 For any $m \times n$ matrix $\mathbf{A}$,

(i) $\text{rk } \mathbf{A} = \text{rk } \mathbf{A}^{\mathrm{T}}$;

(ii) $\text{rk } \mathbf{A} + \text{nul } \mathbf{A} = n$;

(iii) $\text{rk } \mathbf{A} + \text{nul } \mathbf{A}^{\mathrm{T}} = m$.

PROOF If $\mathbf{A} = \mathbf{O}$ then $\mathbf{A}^{\mathrm{T}}$ is also a zero matrix, $\ker \mathbf{A} = R^n$ and $\ker \mathbf{A}^{\mathrm{T}} = R^m$. So from now on we assume that rk $\mathbf{A} > 0$. Indeed, we shall assume that

$$0 < \text{rk } \mathbf{A} < \min (m, n).$$

The second of these strict inequalities ($<$) will not always hold: but the cases rk $\mathbf{A} = m$ and rk $\mathbf{A} = n$ can be dealt with by minor modifications of the argument, which the reader is encouraged to perform.

Let $r = \text{rk } \mathbf{A}$, $s = m - r$, $t = n - r$. By definition of rank, we may choose r linearly independent columns of $\mathbf{A}$. If we rearrange the columns of $\mathbf{A}$ then rng $\mathbf{A}$ and ker $\mathbf{A}^{\mathrm{T}}$ remain the same: the spaces rng $\mathbf{A}^{\mathrm{T}}$ and ker $\mathbf{A}$ are changed by reordering of components, but this does not change their dimensions. We may therefore assume for the purposes of the proof that the *first* r columns of $\mathbf{A}$ are linearly independent.

Suppose now that we apply Gaussian elimination to $\mathbf{A}$. Even though $\mathbf{A}$ is not necessarily square, elimination steps and row interchanges preserve linear independence of columns for the same reason that they preserve nonsingularity in the square case. We may therefore perform r elimination steps on $\mathbf{A}$. This yields the matrix

$$\mathbf{GA} = \begin{pmatrix} \mathbf{U} & \mathbf{W} \\ \mathbf{O}_1 & \mathbf{O}_2 \end{pmatrix}$$

where $\mathbf{G}$ $(m \times m)$ is a row operator,
$\mathbf{U}$ $(r \times r)$ is upper triangular and nonsingular,
$\mathbf{W}$ $(r \times t)$ has no special properties,
$\mathbf{O}_1$ $(s \times r)$ is a zero matrix,
$\mathbf{O}_2$ $(s \times t)$ is a zero matrix.

The only thing that may not be crystal clear about this development is the fact that $\mathbf{O}_2$ is a zero matrix. The first r columns of $\mathbf{A}$ form a basis for rng $\mathbf{A}$. Thus the last t columns of $\mathbf{A}$ are linear combinations of the first r. By the associative law, $\mathbf{GA}$ has the same property. Thus the columns of $\mathbf{O}_2$ are linear combinations of the columns of $\mathbf{O}_1$, which are zeroes by elimination.

(i) Since $\mathbf{G}$ is nonsingular, so is $\mathbf{G}^{\mathrm{T}}$ (Theorem 2.2.4). Thus any m-vector $\mathbf{y}$ may be written in the form $(\mathbf{G}^{\mathrm{T}})^{-1}\mathbf{p}$ where $\mathbf{p} = \mathbf{G}^{\mathrm{T}}\mathbf{y}$. Conversely, any m-vector $\mathbf{p}$ may be written in the form $\mathbf{G}^{\mathrm{T}}\mathbf{y}$ where $\mathbf{y} = (\mathbf{G}^{\mathrm{T}})^{-1}\mathbf{p}$. These facts imply that $\mathbf{A}^{\mathrm{T}}$ has the same range as the matrix $\mathbf{A}^{\mathrm{T}}\mathbf{G}^{\mathrm{T}}$. Now the last s columns of $(\mathbf{GA})^{\mathrm{T}}$ are zeroes; let its first r columns form the matrix $\mathbf{F}$. Then rng $\mathbf{A}^{\mathrm{T}} =$ rng $\mathbf{F}$. But $\mathbf{F}$ is an $n \times r$ matrix whose first r rows form the nonsingular lower triangular matrix $\mathbf{U}^{\mathrm{T}}$, so the columns of $\mathbf{F}$ are linearly independent. Thus $\mathbf{F}$ is a basis for its range, and rk $\mathbf{A}^{\mathrm{T}} = r$.

(ii) Choose any n-vector $\mathbf{x}$ and write it $(\mathbf{x}_1 : \mathbf{x}_2)$ where $\mathbf{x}_1$ denotes the first r components. Recall that $\mathbf{G}$ and $\mathbf{U}$ are nonsingular matrices of orders m and r respectively. Thus

$$\mathbf{Ax} = \mathbf{0},\ \mathbf{GAx} = \mathbf{0},\ \mathbf{Ux}_1 + \mathbf{Wx}_2 = \mathbf{0} \quad \text{and} \quad \mathbf{x}_1 = -\mathbf{U}^{-1}\mathbf{Wx}_2$$

are four equivalent statements. Considering the first and last, we see that the kernel of $\mathbf{A}$ is the range of the matrix

$$\begin{pmatrix} \mathbf{U}^{-1}\mathbf{W} \\ -\mathbf{I}_t \end{pmatrix}$$

The columns of this matrix are linearly independent, so it is a basis for its range. Thus nul $\mathbf{A} = t$.

(iii) The task of proving this by constructing a basis for ker $\mathbf{A}^{\mathrm{T}}$ is left as a challenge to the reader. We shall take a short cut. We know from (ii) that for *any* matrix,

$$\text{rank} + \text{nullity} = \text{number of columns.}$$

In particular, rk $\mathbf{A}^{\mathrm{T}} +$ nul $\mathbf{A}^{\mathrm{T}} = m$ and (iii) follows from (i). QED

Part (i) of Theorem 1 says that the rank of a matrix is its maximal number of linearly independent *rows*. This suggests that there should be some way of characterising rank which is symmetrical in rows and columns. This is done as follows. We define a *submatrix* of $\mathbf{A}$ to be a matrix obtained from $\mathbf{A}$ by deleting some (or none) of its rows and some (or none) of its columns. For example,

$$\begin{pmatrix} 7 & 8 \\ 3 & 2 \end{pmatrix}, \quad \begin{pmatrix} 1 & 2 & 4 \\ 5 & 6 & 8 \\ 9 & 5 & 2 \end{pmatrix}, \quad \begin{pmatrix} 1 & 3 \\ 5 & 7 \\ 9 & 3 \end{pmatrix}$$

are submatrices of

$$\begin{pmatrix} 1 & 2 & 3 & 4 \\ 5 & 6 & 7 & 8 \\ 9 & 5 & 3 & 2 \end{pmatrix}.$$

THEOREM 2 The rank of a matrix is the maximal order of its nonsingular submatrices.

PROOF Let $\mathbf{A}$ be an $m \times n$ matrix of rank r. If $r = 0$ there is nothing to prove. So assume $r > 0$. Since r is the dimension of rng $\mathbf{A}$, any $r + 1$ columns of $\mathbf{A}$ are linearly dependent: thus any square submatrix of order greater than r is singular. It remains to show that $\mathbf{A}$ has a nonsingular $r \times r$ submatrix.

Select r linearly independent columns of $\mathbf{A}$: let the matrix so formed be $\mathbf{B}$. By Theorem 1, $\mathbf{B}^{\mathrm{T}}$ is an $r \times r$ matrix with the same rank as $\mathbf{B}$, namely r. Select r linearly independent columns of $\mathbf{B}^{\mathrm{T}}$: let the matrix so formed be $\mathbf{C}$. Then $\mathbf{C}$ is $r \times r$ nonsingular and so is $\mathbf{C}^{\mathrm{T}}$. Now $\mathbf{C}$ is a submatrix of $\mathbf{B}^{\mathrm{T}}$, so $\mathbf{C}^{\mathrm{T}}$ is a submatrix of $\mathbf{B}$ and therefore of $\mathbf{A}$. QED

We now state and prove the two theorems on solutions of systems of linear equations. The first uses part (ii) of Theorem 1, which we write down again in mnemonic form:

(ii) rank + nullity = number of columns.

THEOREM 3 Let $\mathbf{A}$ be an $m \times n$ matrix, $\mathbf{b}$ an m-vector. Let r be the rank of $\mathbf{A}$ and let r^* be the rank of the augmented matrix $(\mathbf{A}, \mathbf{b})$. Let S denote the set of all solutions $\mathbf{x}$ to the system $\mathbf{Ax} = \mathbf{b}$.

(I) If $r^* = r + 1$, $S = \varnothing$: no solution.

(II) If $r^* = r = n$, S is a singleton: unique solution.

(III) If $r^* = r < n$ there are infinitely many solutions. In this case S consists of all n-vectors of the form $\mathbf{x}^0 + \mathbf{z}$, where $\mathbf{x}^0$ is a particular solution and $\mathbf{z}$ is an arbitrary member of the $(n - r)$-dimensional kernel of $\mathbf{A}$.

These are the only possible cases.

PROOF We prove the last statement first. Since **A** has n columns, $r \leq n$. Augmenting **A** by **b** can expand rank by at most 1. Thus r^* is either r or $r + 1$.

By the First Law of Linear Dependence, $r^* = r$ if and only if **b** is in the range of **A**. This proves (I) and also shows that S is non-empty when $r^* = r$. The other parts of (II) and (III) now follow from (ii) and the fact that the only zero-dimensional subspace of R^n is $\{\mathbf{0}\}$. QED

When **A** is square, Theorem 3 reduces to the properties of square systems which we summarised at the beginning of the last section. Let $m = n$. If **A** is nonsingular then $r = n$ and $r^* = n$ for any **b**, so a unique solution always exists (II). If **A** is singular then $r < n$: here r^* may be either r or $r + 1$ depending on whether **b** is or is not in the range of **A**; in the former case there are infinitely many solutions (III) and in the latter no solution exists (I).

The other theorem on systems of linear equations is the Theorem of the Alternative, which will provide the foundation of our discussion of orthogonality in Chapter 4 and will be extended to systems of linear inequalities in Section 8.3. The proof of this theorem uses part (iii) of Theorem 1:

(iii) rank + nul (transpose) = number of rows.

THEOREM 4 (Theorem of the Alternative)
Let **A** be an $m \times n$ matrix, **b** an m-vector. Then *either* there exists an n-vector **x** such that $\mathbf{Ax} = \mathbf{b}$ *or* there exists an m-vector **p** such that $\mathbf{A}^{\mathrm{T}}\mathbf{p} = \mathbf{0}$ and $\mathbf{b}^{\mathrm{T}}\mathbf{p} \neq 0$ *but not both.*

PROOF If $\mathbf{Ax} = \mathbf{b}$ and $\mathbf{A}^{\mathrm{T}}\mathbf{p} = \mathbf{0}$ then $\mathbf{b}^{\mathrm{T}}\mathbf{p} = \mathbf{x}^{\mathrm{T}}\mathbf{A}^{\mathrm{T}}\mathbf{p} = 0$. This proves 'but not both'. To prove 'either/or' suppose that the system $\mathbf{Ax} = \mathbf{b}$ has no solution and let **C** denote the augmented matrix (**A**, **b**). By Theorem 3,

$$\text{rk } \mathbf{C} = \text{rk } \mathbf{A} + 1.$$

Subtracting both sides from m and applying (iii),

$$\text{nul } \mathbf{C}^{\mathrm{T}} = \text{nul } \mathbf{A}^{\mathrm{T}} - 1.$$

We may therefore choose an m-vector **p** which is in ker $\mathbf{A}^{\mathrm{T}}$ but not in ker $\mathbf{C}^{\mathrm{T}}$. Hence $\mathbf{A}^{\mathrm{T}}\mathbf{p} = \mathbf{0}$ but $\mathbf{b}^{\mathrm{T}}\mathbf{p} \neq 0$. QED

We now give a general solution procedure for the $m \times n$ system $\mathbf{Ax} = \mathbf{b}$. Suppose first that rk $\mathbf{A} = r$ and the first r columns of $\mathbf{A}$ are linearly independent. We proceed by Gaussian elimination on the augmented matrix $(\mathbf{A}, \mathbf{b})$. After r elimination steps, possibly accompanied by row interchanges, we obtain the matrix

$$\begin{pmatrix} \mathbf{U} & \mathbf{W} & \mathbf{d}_1 \\ \mathbf{O}_1 & \mathbf{O}_2 & \mathbf{d}_2 \end{pmatrix}.$$

(Notice that $\mathbf{d}_2$ and the zero matrices $\mathbf{O}_1$ and $\mathbf{O}_2$ are not present if $r = m$, while $\mathbf{W}$ and $\mathbf{O}_2$ are not present if $r = n$.) Letting $\mathbf{x} = (\mathbf{x}_1 : \mathbf{x}_2)$, where $\mathbf{x}_1$ consists of the first r components, we have the equivalent system

$$\mathbf{Ux}_1 + \mathbf{Wx}_2 = \mathbf{d}_1, \quad \mathbf{0} = \mathbf{d}_2 .$$

Now $\mathbf{d}_2$ is a vector of *constants* depending on $\mathbf{A}$ and $\mathbf{b}$. So if $\mathbf{d}_2$ is not in fact a zero vector, the system $\mathbf{Ax} = \mathbf{b}$ is inconsistent: there is no solution. If $\mathbf{d}_2 = \mathbf{0}$ the set of all solutions is found by assigning arbitrary values to the components of $\mathbf{x}_2$ and setting

$$\mathbf{x}_1 = \mathbf{U}^{-1}(\mathbf{d}_1 - \mathbf{Wx}_2).$$

Since $\mathbf{U}$ is upper triangular, this last step can be accomplished by the familiar process of back-substitution.

The argument just given does not assume that we know the rank, r, of $\mathbf{A}$ before we start, but it does assume that the *first* r columns of $\mathbf{A}$ are linearly independent. This cannot be guaranteed, but it is easy to deal with the case where it does not hold: if we find that elimination on a given column is impossible, we perform a *column interchange*. The only tricky part of this is making sure that we solve for the right variables when the time comes for back-substitution. We therefore start by labelling the columns of $\mathbf{A}$ and interchange labels when we interchange columns.

Example 1 Find all solutions to the system of equations

$$\begin{aligned} x_1 - 2x_2 + x_3 + x_4 + 2x_5 &= 2 \\ 3x_1 - 6x_2 + 5x_3 + 2x_4 + x_5 &= 3 \\ -x_1 + 2x_2 - 5x_3 + x_4 + 8x_5 &= 4 \\ 2x_1 - 4x_2 - 6x_3 - 5x_4 + 2x_5 &= 5. \end{aligned}$$

After writing the augmented matrix, labelling the columns and performing one elimination step we have

1	2	3	4	5	
1	−2	1	1	2	2
0	0	2	−1	−5	−3
0	0	−4	2	10	6
0	0	−8	−7	−2	1

We need a column interchange before the next elimination step. Since the column labelled 2 is of no further use to us, we move it over to the right, shifting columns 3, 4, 5 one place to the left. This, followed by an elimination step, yields

1	3	4	5	2	
1	1	1	2	−2	2
0	2	−1	−5	0	−3
0	0	0	0	0	0
0	0	−11	−22	0	−11

A row interchange, which we do not bother to perform, yields the final form of the system. Notice that the equation with zero coefficients has zero on the right-hand side, so the system is consistent. Assigning arbitrary values λ and 2μ to x_2 and x_5 respectively, we can solve for x_4, x_3 and x_1 by back-substitution:

$$x_4 = 1 - 4\mu,$$
$$x_3 = (x_4 + 10\mu - 3)/2 = 3\mu - 1,$$
$$x_1 = 2 + 2\lambda - 4\mu - (x_3 + x_4) = 2 + 2\lambda - 3\mu.$$

Thus the general solution is

$$x_1 = 2 + 2\lambda - 3\mu, \quad x_2 = \lambda, \; x_3 = 3\mu - 1,$$

$$x_4 = 1 - 4\mu, \quad x_5 = 2\mu$$

where λ and μ are arbitrary numbers.

Example 2 Consider the system which differs from that of Example 1 in only one respect: the third equation is now

$$-x_1 + 2x_2 - 5x_3 + 2x_4 + 10x_5 = 9.$$

We show that this system has no solution.

The table after two elimination steps is now

1	3	4	5	2	
1	1	1	2	−2	2
0	2	−1	−5	0	−3
0	0	1	2	0	5
0	0	−11	−22	0	−11

One more elimination step gives

1	3	4	5	2	
1	1	1	2	−2	2
0	2	−1	−5	0	−3
0	0	1	2	0	5
0	0	0	0	0	44

The last equation reads '$0 = 44$' which reveals an inconsistency.

We end this section with some more inequalities concerning rank.

THEOREM 5 Let $\mathbf{C} = \mathbf{AB}$. Then

(i) $\text{rk}\,\mathbf{C} \leq \text{rk}\,\mathbf{B}$ with equality if the columns of $\mathbf{A}$ are linearly independent;

(ii) $\text{rk}\,\mathbf{C} \leq \text{rk}\,\mathbf{A}$ with equality if the rows of $\mathbf{B}$ are linearly independent.

PROOF $\mathbf{C}^{\mathrm{T}} = \mathbf{B}^{\mathrm{T}}\mathbf{A}^{\mathrm{T}}$: also $\text{rk}\,\mathbf{A} = \text{rk}\,\mathbf{A}^{\mathrm{T}}$, and similarly for $\mathbf{B}$ and $\mathbf{C}$. Thus we need only prove one of (i) and (ii), and we prove the former. Let (π) stand for the phrase 'with equality if the columns of $\mathbf{A}$ are linearly independent'. Then

$$\ker \mathbf{B} \subset \ker \mathbf{C}. \quad (\pi)$$

Taking dimensions,

$$\text{nul}\,\mathbf{B} \leq \text{nul}\,\mathbf{C}. \quad (\pi)$$

But $\mathbf{B}$ and $\mathbf{C}$ have the same number of columns. Thus

$$\text{rk}\,\mathbf{B} \geq \text{rk}\,\mathbf{C} \qquad (\pi)$$

QED

The inequalities in Theorem 5 can be iterated: thus if the matrix $\mathbf{P}$ can be expressed as a product of matrices, one of which is $\mathbf{A}$, then $\text{rk}\,\mathbf{P} \leq \text{rk}\,\mathbf{A}$.

THEOREM 6 If $\mathbf{A}$ and $\mathbf{B}$ are $m \times n$ matrices, then $\text{rk}(\mathbf{A}+\mathbf{B}) \leq \text{rk}\ \mathbf{A} + \text{rk}\ \mathbf{B}$.

PROOF Let $r = \text{rk}\ \mathbf{A} + \text{rk}\ \mathbf{B}$. It suffices to express $\mathbf{A}+\mathbf{B}$ as a product of matrices, one of which has rank r. Letting

$$\mathbf{C} = (\mathbf{I}_m \quad \mathbf{I}_m), \quad \mathbf{D} = \begin{pmatrix} \mathbf{A} & \mathbf{O} \\ \mathbf{O} & \mathbf{B} \end{pmatrix}, \quad \mathbf{E} = \begin{pmatrix} \mathbf{I}_n \\ \mathbf{I}_n \end{pmatrix}$$

we have $\mathbf{A}+\mathbf{B} = \mathbf{CDE}$ and $\text{rk}\ \mathbf{D} = r$. QED

Exercises

1. Find all solutions to the following systems of equations.

   (i) $$\begin{aligned} x_1 + x_2 + x_3 + x_4 &= 1 \\ x_1 + 2x_2 + 3x_3 + 4x_4 &= 2 \end{aligned}$$

   (ii) $$\begin{aligned} 2x_1 + x_2 - x_3 &= 7 \\ 4x_1 + 2x_2 - 3x_3 &= 8 \\ 3x_1 + 2x_2 - 2x_3 &= 10 \\ 5x_1 - 6x_2 + 4x_3 &= 14 \end{aligned}$$

   (iii) $$\begin{aligned} x_1 + 2x_2 + 3x_3 &= 15 \\ 4x_1 + 5x_2 + 6x_3 &= 30 \\ 7x_1 + 8x_2 + 9x_3 &= 45 \end{aligned}$$

2. Illustrate the Theorem of the Alternative using the data of Example 2.

3. Show that the matrix

$$\begin{pmatrix} 0 & 2 & 1 & 1 \\ 1 & 0 & 2 & 2 \\ 3 & 2 & 4 & 7 \\ 2 & 0 & 1 & 4 \end{pmatrix}$$

   is singular and find bases for its range and kernel.

4. Prove that, if $\mathbf{B}$ is an $n \times r$ matrix of rank r, there exists a nonsingular matrix $\mathbf{A}$ whose first r columns form $\mathbf{B}$.

3 Pivoting

We have yet to give a detailed recipe for inverting a matrix. Section 3.1 does this, using the method of Gauss–Jordan pivoting. This method can be given a condensed computational layout which is rather easier to work with than the usual one. The use of the condensed form is sometimes referred to as modified Jordan elimination, and its leading modern exponent is A. W. Tucker: we refer to it as Tucker pivoting and explain it in Section 3.2.

Tucker pivoting has applications far beyond the inversion of square matrices. It leads to a device called the method of complementary solutions. Following Tucker (1974), we shall use this device as our main tool in linear programming. The method is explained in Section 3.3, but the motivation for using it is not revealed until the end of Chapter 5.

We make no apology for keeping the reader in the dark, since we feel that he should master the arithmetical trick used in linear programming before facing the difficulties inherent in the subject itself. Thus the reader is strongly encouraged to read Section 3.3 before beginning Chapter 5.

One point of notation before we begin. In this chapter, r stands for 'row', not for 'rank'. Thus there is nothing special about the 'rth elimination step'.

3.1 Gauss–Jordan pivoting

Let $\mathbf{A}$ be a nonsingular matrix of order n. Consider the equation system

$$\mathbf{Ax} + \mathbf{y} = \mathbf{0} \tag{1}$$

in $2n$ variables $x_1, \ldots, x_n, y_1, \ldots, y_n$. Premultiplying by $\mathbf{A}^{-1}$ we have

$$\mathbf{x} + \mathbf{A}^{-1}\mathbf{y} = \mathbf{0}. \tag{2}$$

If we can get from (1) to the equivalent system (2) in a sequence of steps, we shall have revealed $\mathbf{A}^{-1}$. That is the main task of this section. (Of course we could work equally well with the system $\mathbf{Ax} = \mathbf{y}$; the reason for the sign convention of (1) is that it fits in better with our subsequent discussion.)

Let us see how far we can get with Gaussian elimination, starting with the $n \times 2n$ matrix $(\mathbf{A}, \mathbf{I})$. When elimination is complete we have the system

$$\mathbf{Ux} + \mathbf{Gy} = \mathbf{0} \tag{3}$$

where $\mathbf{U}$ is upper triangular and $\mathbf{G}$ is a row operator. Since $\mathbf{U}$ is not in general the identity matrix or even diagonal, (3) is not identical to (2). But this argument suggests a variant of Gaussian elimination which will do the trick: eliminate above the diagonal as well as below!

Recall that the rth Gaussian elimination step is:

$$\begin{aligned} \beta_{ij} &= \alpha_{ij} - \alpha_{ir}\alpha_{rr}^{-1}\alpha_{ij}, && \text{for } i > r \text{ and all } j, \\ \beta_{ij} &= \alpha_{ij}, && \text{for } i \leq r \text{ and all } j. \end{aligned}$$

The Gauss–Jordan variant is

$$\begin{aligned} \beta_{ij} &= \alpha_{ij} - \alpha_{ir}\alpha_{rr}^{-1}\alpha_{rj}, && \text{for } i \neq r \text{ and all } j, \\ \beta_{rj} &= \alpha_{rj}, && \text{for all } j. \end{aligned}$$

We call this the rth *pivot step* with α_{rr} as *pivot*. After n pivot steps, possibly interspersed with row interchanges, we have the system

$$\mathbf{Dx} + \mathbf{Hy} = \mathbf{0} \tag{4}$$

where $\mathbf{D}$ is a nonsingular *diagonal* matrix and $\mathbf{H}$ is nonsingular. Comparing (4) with (2) we see that $\mathbf{A}^{-1} = \mathbf{D}^{-1}\mathbf{H}$. Since $\mathbf{D}$ is diagonal, the procedure of obtaining $\mathbf{A}^{-1}$ from $\mathbf{H}$ and $\mathbf{D}$ is simply a matter of dividing each row of $\mathbf{H}$ by the relevant entry of $\mathbf{D}$.

Example 1 Invert the matrix

$$\mathbf{A} = \begin{pmatrix} 2 & 1 & 1 \\ 4 & 2 & 5 \\ 3 & 2 & 4 \end{pmatrix}.$$

We start by writing the 3×6 matrix $(\mathbf{A}, \mathbf{I})$

$$\begin{array}{ccc|ccc} 2 & 1 & 1 & 1 & 0 & 0 \\ 4 & 2 & 5 & 0 & 1 & 0 \\ 3 & 2 & 4 & 0 & 0 & 1 \end{array}$$

The first pivot step is as in Gaussian elimination:

$$\begin{array}{ccc|ccc} 2 & 1 & 1 & 1 & 0 & 0 \\ 0 & 0 & 3 & -2 & 1 & 0 \\ 0 & \frac{1}{2} & \frac{5}{2} & -\frac{3}{2} & 0 & 1 \end{array}$$

Now we need a row interchange:

$$\begin{array}{ccc|ccc} 2 & 1 & 1 & 1 & 0 & 0 \\ 0 & \frac{1}{2} & \frac{5}{2} & -\frac{3}{2} & 0 & 1 \\ 0 & 0 & 3 & -2 & 1 & 0 \end{array}$$

The second pivot step marks our first departure from Gaussian elimination: subtract twice the second row from the first. The third row remains unchanged as there is already a zero (3, 2) entry.

$$\begin{array}{ccc|ccc} 2 & 0 & -4 & 4 & 0 & -2 \\ 0 & \frac{1}{2} & \frac{5}{2} & -\frac{3}{2} & 0 & 1 \\ 0 & 0 & 3 & -2 & 1 & 0 \end{array}$$

In the third and last pivot step we add $\frac{4}{3}$ times the third row to the first and subtract $\frac{5}{6}$ times the third row from the second. This gives our $(\mathbf{D}, \mathbf{H})$ form:

$$\begin{array}{ccc|ccc} 2 & 0 & 0 & \frac{4}{3} & \frac{4}{3} & -2 \\ 0 & \frac{1}{2} & 0 & \frac{1}{6} & -\frac{5}{6} & 1 \\ 0 & 0 & 3 & -2 & 1 & 0 \end{array}$$

The inverse is obtained by dividing each row of $\mathbf{H}$ by the relevant entry of $\mathbf{D}$:

$$\mathbf{A}^{-1} = \begin{pmatrix} \frac{2}{3} & \frac{2}{3} & -1 \\ \frac{1}{3} & -\frac{5}{3} & 2 \\ -\frac{2}{3} & \frac{1}{3} & 0 \end{pmatrix}.$$

This section began with the words 'let **A** be a nonsingular matrix'. But the operations on and below the diagonal of **A** are the same in Gauss–Jordan pivoting as in Gaussian elimination, so the same singularity test is applicable to both methods.

There is a variant of Gauss–Jordan pivoting which in itself is not particularly convenient, but acts as a prelude to the next section. In the procedure described above, we obtained $(\mathbf{D}, \mathbf{H})$ and *then* found $\mathbf{A}^{-1}$ by division. We call the variant DAWG, which stands for Divide As We Go. The rth pivot step in DAWG is

$$\begin{aligned} \beta_{ij} &= \alpha_{ij} - \alpha_{ir}\alpha_{rr}^{-1}\alpha_{rj}, && \text{for } i \neq r \text{ and all } j,\\ \beta_{rj} &= \alpha_{rj}/\alpha_{rr}, && \text{for all } j. \end{aligned}$$

After n of these steps (and possibly some row interchanges) we obtain $(\mathbf{I}, \mathbf{A}^{-1})$.

Suppose now that we want to find $\mathbf{A}^{-1}$ *and* solve the system $\mathbf{Ax} = \mathbf{b}$ for a particular $\mathbf{b}$. One way to do this is to compute $\mathbf{A}^{-1}$ and postmultiply it by $\mathbf{b}$. A slightly quicker way is to bring $\mathbf{b}$ into the pivoting procedure from the start. We begin with $(\mathbf{A} \quad \mathbf{I} \quad \mathbf{b})$ and end with $(\mathbf{I} \quad \mathbf{A}^{-1} \quad \mathbf{A}^{-1}\mathbf{b})$. The next example implements this procedure with DAWG, but clearly it can be done with final division instead.

Example 2

$$\text{Given} \quad \mathbf{A} = \begin{pmatrix} 3 & 2 & 1 \\ 1 & 1 & 4 \\ 1 & 1 & 5 \end{pmatrix}, \quad \mathbf{b} = \begin{pmatrix} 6 \\ 8 \\ 9 \end{pmatrix}$$

find $\mathbf{A}^{-1}$ and $\mathbf{A}^{-1}\mathbf{b}$.

We start with $(\mathbf{A} \quad \mathbf{I} \quad \mathbf{b})$

$$\left(\begin{array}{ccc|ccc|c} 3 & 2 & 1 & 1 & 0 & 0 & 6 \\ 1 & 1 & 4 & 0 & 1 & 0 & 8 \\ 1 & 1 & 5 & 0 & 0 & 1 & 9 \end{array}\right)$$

In the first DAWG step we subtract $\frac{1}{3}$ times the first row from each of the other two rows *and then divide* the first row by 3.

$$\left(\begin{array}{ccc|ccc|c} 1 & \frac{2}{3} & \frac{1}{3} & \frac{1}{3} & 0 & 0 & 2 \\ 0 & \frac{1}{3} & \frac{11}{3} & -\frac{1}{3} & 1 & 0 & 6 \\ 0 & \frac{1}{3} & \frac{14}{3} & -\frac{1}{3} & 0 & 1 & 7 \end{array}\right)$$

The next two DAWG steps yield the following arrays

$$\begin{array}{rrr|rrr|r} 1 & 0 & -7 & 1 & -2 & 0 & -10 \\ 0 & 1 & 11 & -1 & 3 & 0 & 18 \\ 0 & 0 & 1 & 0 & -1 & 1 & 1 \end{array}$$

$$\begin{array}{rrr|rrr|r} 1 & 0 & 0 & 1 & -9 & 7 & -3 \\ 0 & 1 & 0 & -1 & 14 & -11 & 7 \\ 0 & 0 & 1 & 0 & -1 & 1 & 1 \end{array}$$

Hence

$$\mathbf{A}^{-1} = \begin{pmatrix} 1 & -9 & 7 \\ -1 & 14 & -11 \\ 0 & -1 & 1 \end{pmatrix} \text{ and } \mathbf{A}^{-1}\mathbf{b} = \begin{pmatrix} -3 \\ 7 \\ 1 \end{pmatrix}.$$

Exercises

1. Invert the matrix

$$\begin{pmatrix} 0 & 3 & -2 \\ 2 & 5 & 3 \\ 1 & 4 & 1 \end{pmatrix}.$$

2. Let

$$\mathbf{A} = \begin{pmatrix} 0 & 1 & 3 \\ 2 & 0 & 1 \\ 1 & 4 & 0 \end{pmatrix}, \quad \mathbf{b} = \begin{pmatrix} 8 \\ 0 \\ 7 \end{pmatrix}.$$

Find $\mathbf{A}$ and $\mathbf{A}^{-1}\mathbf{b}$.

3.2 Tucker pivoting

Let $\mathbf{A}$ be a nonsingular matrix of order n, $\mathbf{b}$ an n-vector. Suppose we calculate $\mathbf{A}^{-1}$ and $\mathbf{A}^{-1}\mathbf{b}$ using DAWG. Each array that we write is of the form

$$\mathbf{S} = (\mathbf{GA} \quad \mathbf{G} \quad \mathbf{Gb}),$$

where $\mathbf{G}$ is a nonsingular matrix. Further, it is in the nature of DAWG that n of the first $2n$ columns of $\mathbf{S}$ are a rearrangement of the columns of $\mathbf{I}_n$.

Let us rephrase this in terms of equation systems. The system we start with is

$$\mathbf{A}\mathbf{x} + \mathbf{y} = \mathbf{b}. \tag{1}$$

The array **S** depicts an equivalent system of the form

$$\tilde{\mathbf{A}}\tilde{\mathbf{x}} + \tilde{\mathbf{y}} = \tilde{\mathbf{b}} \tag{2}$$

where $(\tilde{\mathbf{x}} : \tilde{\mathbf{y}})$ is a rearrangement of the components of $(\mathbf{x} : \mathbf{y})$. Let $\mathbf{z}$ and $\tilde{\mathbf{z}}$ denote the $2n$-vectors $(\mathbf{x} : \mathbf{y})$ and $(\tilde{\mathbf{x}} : \tilde{\mathbf{y}})$ respectively: then

$$\tilde{z}_k = z_{\pi k} \quad \text{for } k = 1, \ldots, 2n$$

where $(\pi_1, \ldots, \pi_{2n})$ is a *permutation* (in other words, a rearrangement) of the list of integers $1, \ldots, 2n$.

Observe that the only information we need to describe the system (2) in an array is the matrix $\tilde{\mathbf{A}}$, the vector $\tilde{\mathbf{b}}$ and the *labels* $\pi_1, \ldots, \pi_{2n}$. So instead of writing down a Gauss–Jordan array (a matrix with n rows and $2n + 1$ columns) we may write the *Tucker scheme*

$$\begin{array}{c|c|c} & & \pi_1 \ \cdots \ \pi_n \\ \hline \pi_{n+1} & & \\ \vdots & \tilde{\mathbf{b}} & \tilde{\mathbf{A}} \\ \pi_{2n} & & \end{array}$$

The initial scheme, describing (1), is

$$\begin{array}{c|c|c} & & 1 \ \cdots \ n \\ \hline n+1 & & \\ \vdots & \mathbf{b} & \mathbf{A} \\ 2n & & \end{array}$$

The final scheme, from which we may read $\mathbf{A}^{-1}$ and $\mathbf{A}^{-1}\mathbf{b}$, is

$$\begin{array}{c|c|c} & & n+1 \ \cdots \ 2n \\ \hline 1 & & \\ \vdots & \mathbf{A}^{-1}\mathbf{b} & \mathbf{A}^{-1} \\ n & & \end{array}$$

What we must do now is to interpret the DAWG pivot step as a method of going from one scheme to another. Let us first consider cases (as in Example 3.1.2) where the solution is reached by pivot steps alone, without row interchanges. At the rth pivot step, we must make the old $\tilde{x}_r$ into the new $\tilde{y}_r$, and vice versa. This involves three operations.

(i) Perform a DAWG step with the (r, r) entry of $\tilde{\mathbf{A}}$ as pivot. This reduces the column labelled r to the rth column of $\mathbf{I}_n$.
(ii) Replace this uninteresting column by what would have been column $n + r$ of the Gauss–Jordan array.
(iii) Interchange the labels of column r and row r.

We call operations (i) and (ii) a *Tucker pivot* on the rth diagonal entry. We call (iii) a *switch*. The whole step is called a *pivot-and-switch* (PS) operation and is described algebraically as follows:

$$
\begin{aligned}
\text{PIVOT: } \beta_{ij} &= \alpha_{ij} - \alpha_{ir}\alpha_{rr}^{-1}\alpha_{rj}, &&\text{for } i \neq r \text{ and } j \neq r,\\
\beta_{rj} &= \alpha_{rj}/\alpha_{rr}, &&\text{for } j \neq r,\\
\beta_{ir} &= -\alpha_{ir}/\alpha_{rr}, &&\text{for } i \neq r,\\
\beta_{rr} &= 1/\alpha_{rr}.
\end{aligned}
$$

SWITCH: Interchange labels π_r and π_{n+r}.

We put this procedure into practice by repeating Example 3.1.2 using Tucker rather than Gauss–Jordan. It is usual to circle the pivot entry.

Example 1 Given

$$
\mathbf{A} = \begin{pmatrix} 3 & 2 & 1 \\ 1 & 1 & 4 \\ 1 & 1 & 5 \end{pmatrix}, \quad \mathbf{b} = \begin{pmatrix} 6 \\ 8 \\ 9 \end{pmatrix},
$$

find $\mathbf{A}^{-1}$ and $\mathbf{A}^{-1}\mathbf{b}$.

The initial scheme is

		1	2	3
4	6	(3)	2	1
5	8	1	1	4
6	9	1	1	5

The first PS operation interchanges the labels 4 and 1. Also, since the pivot entry is 3 and $\frac{1}{3}$ times the first equation is being subtracted from each of the other equations, the column labelled 4 in the new scheme should be $(\frac{1}{3}, -\frac{1}{3}, -\frac{1}{3})^{\mathrm{T}}$. Everything else is as in DAWG.

		4	2	3
1	2	$\frac{1}{3}$	$\frac{2}{3}$	$\frac{1}{3}$
5	6	$-\frac{1}{3}$	($\frac{1}{3}$)	$\frac{11}{3}$
6	7	$-\frac{1}{3}$	$\frac{1}{3}$	$\frac{14}{3}$

The next two schemes derived by PS operations are as follows:

		4	5	3
1	−10	1	−2	−7
2	18	−1	3	11
6	1	0	−1	(1)

		4	5	6
1	−3	1	−9	7
2	7	−1	14	−11
3	1	0	−1	1

From the last scheme we read off the answer:

$$\mathbf{A}^{-1} = \begin{pmatrix} 1 & -9 & -7 \\ -1 & 14 & -11 \\ 0 & -1 & 1 \end{pmatrix}, \quad \mathbf{A}^{-1}\mathbf{b} = \begin{pmatrix} -3 \\ 7 \\ 1 \end{pmatrix}.$$

An obvious advantage of the Tucker formulation of Gauss–Jordan is that we do not need to write down the columns of the identity matrix at each step. Notice also that we can avoid row interchanges. There are two aspects to this. First, we can pivot on diagonal entries *in any order*. Second, we can pivot on non-diagonal entries if we so wish. All we are really after is a scheme of the form

		π_1 ⋯ π_n
π_{n+1} ⋮ π_{2n}	$\tilde{\mathbf{b}}$	$\tilde{\mathbf{A}}$

where $(\pi_1, \ldots, \pi_n)$ is some permutation of the big labels $(n+1, \ldots, 2n)$ and $(\pi_{n+1}, \ldots, \pi_{2n})$ is some permutation of the small labels $1, \ldots, n$). Then the (i, j) entry of $\mathbf{A}^{-1}$ is the entry of $\tilde{\mathbf{A}}$ in the row *labelled* i and the column *labelled* $n + j$. The ith component of $\mathbf{A}^{-1}\mathbf{b}$ is the component of $\tilde{\mathbf{b}}$ in the row *labelled* i.

Since we are now allowing non-diagonal pivots we need a new recipe. Suppose we pivot on the entry α_{rs} of a scheme whose entries are (α_{ij}) and whose labels are (π_k). Suppose that this yields a new scheme whose entries are (β_{ij}) and whose labels are (ψ_k). Then the full description of the PS operation is

$$
\begin{aligned}
&\beta_{ij} = \alpha_{ij} - \alpha_{is}\alpha_{rs}^{-1}\alpha_{rj}, && \text{for } i \neq r \text{ and } j \neq s,\\
&\beta_{rj} = \alpha_{rj}/\alpha_{rs}, && \text{for } j \neq s,\\
&\beta_{is} = -\alpha_{is}/\alpha_{rs}, && \text{for } i \neq r,\\
&\beta_{rs} = 1/\alpha_{rs}, &&\\
&\psi_s = \pi_{n+r}, \quad \psi_{n+r} = \pi_s \quad \text{and} \quad \psi_k = \pi_k \text{ for all other } k. &&
\end{aligned}
$$

Example 2 Invert the matrix

$$
\mathbf{A} = \begin{pmatrix} 0 & 3 & 4 \\ 2 & 0 & 3 \\ -1 & 1 & 0 \end{pmatrix}.
$$

Since we are interested only in inversion, we use schemes without a left border. The initial scheme is

	1	2	3
4	0	3	4
5	2	0	3
6	−1	$\textcircled{1}$	0

We now pivot and switch until all the original column labels have been transferred to rows. One fairly painless sequence is as follows.

	1	6	3
4	3	−3	4
5	$\textcircled{2}$	0	3
2	−1	1	0

	5	6	3
4	$-\frac{3}{2}$	−3	$\textcircled{-\frac{1}{2}}$
1	$\frac{1}{2}$	0	$\frac{3}{2}$
2	$\frac{1}{2}$	1	$\frac{3}{2}$

	5	6	4
3	3	6	−2
1	−4	−9	3
2	−4	−8	3

The inverse is read off using the rule that its (i, j) entry is the entry in the row labelled i and the column labelled $3 + j$ in the last scheme.

Thus

$$\mathbf{A}^{-1} = \begin{pmatrix} 3 & -4 & -9 \\ 3 & -4 & -8 \\ -2 & 3 & 6 \end{pmatrix}.$$

We have worked with schemes with one border and schemes with none. Now we introduce schemes with two borders.

Let $\mathbf{A}$ be a nonsingular matrix of order n, and let $\mathbf{b}$ and $\mathbf{c}$ be n-vectors. We wish to find $\mathbf{A}^{-1}$, $\mathbf{A}^{-1}\mathbf{b}$ and $(\mathbf{A}^{\mathrm{T}})^{-1}\mathbf{c}$. We work with two systems, each involving n equations and $2n$ variables. The familiar system, with variables $x_1, \ldots, x_n, y_1, \ldots, y_n$ is

$$\mathbf{A}\mathbf{x} + \mathbf{y} = \mathbf{b}. \tag{3}$$

The unfamiliar system, with variables $q_1, \ldots, q_n, p_1, \ldots, p_n$ is

$$\mathbf{q} - \mathbf{A}^{\mathrm{T}}\mathbf{p} = -\mathbf{c}. \tag{4}$$

The sign convention in (4) looks quite bizarre but it will make sense eventually!

Suppose now that we premultiply (3) by $\mathbf{A}^{-1}$ and (4) by $(-\mathbf{A}^{\mathrm{T}})^{-1}$. We have the following systems, which are equivalent to (3) and (4) respectively:

$$\mathbf{x} + \mathbf{A}^{-1}\mathbf{y} = \mathbf{A}^{-1}\mathbf{b}, \tag{5}$$

$$-(\mathbf{A}^{\mathrm{T}})^{-1}\mathbf{q} + \mathbf{p} = (\mathbf{A}^{\mathrm{T}})^{-1}\mathbf{c}. \tag{6}$$

If we can pass from (3) and (4) to (5) and (6) in a sequence of steps, we shall reveal what we desire.

The starting point is the following scheme with two borders:

		$1 \;\cdots\; n$
	0	$-\mathbf{c}^{\mathrm{T}}$
$n+1$ ⋮ $2n$	$\mathbf{b}$	$\mathbf{A}$

Apart from the top row, this is quite familiar: the *row system* of the scheme is defined to be (3). We define the *column system* of the scheme to be (4). Notice in particular that the $-\mathbf{c}^{\mathrm{T}}$ in the top row goes with the $-\mathbf{c}$ on the right hand side of (4). The significance of the zero in the top left corner will become clear in the next section.

Now suppose we perform some PS operations: these are applied to the entire $(1 + n) \times (1 + n)$ scheme but only entries in labelled rows and columns are used as pivots. The typical scheme will be of the form

$$
\begin{array}{r|c|ccc}
 & & \pi_1 & \cdots & \pi_n \\
\hline
\Sigma = & \alpha_{00} & \alpha_{01} & \cdots & \alpha_{0n} \\
\hline
\pi_{n+1} & \alpha_{10} & \alpha_{11} & \cdots & \alpha_{1n} \\
\vdots & \vdots & & \vdots & \\
\pi_{2n} & \alpha_{n0} & \alpha_{n1} & \cdots & \alpha_{nn}
\end{array}
$$

The row system of Σ is

$$
\left.\begin{array}{l}
\alpha_{i1}\tilde{z}_1 + \alpha_{i2}\tilde{z}_2 + \cdots + \tilde{\alpha}_{in}\tilde{z}_n + \tilde{z}_{n+i} = \alpha_{i0} \\
\quad \text{for } i = 1, \ldots, n, \text{ where} \\
\tilde{z}_k \text{ is component } \pi_k \text{ of the } 2n\text{-vector } \mathbf{z} = (\mathbf{x} : \mathbf{y}).
\end{array}\right\} \tag{7}
$$

The column system of Σ is

$$
\left.\begin{array}{l}
\tilde{w}_j - \alpha_{ij}\tilde{w}_{n+1} - \alpha_{2j}\tilde{w}_{n+2} - \cdots - \alpha_{nj}\tilde{w}_{2n} = \alpha_{0j} \\
\quad \text{for } j = 1, \ldots, n, \text{ where} \\
\tilde{w}_k \text{ is component } \pi_k \text{ of the } 2n\text{-vector } \mathbf{w} = (\mathbf{q} : \mathbf{p}).
\end{array}\right\} \tag{8}
$$

Our approach to Tucker pivoting via Gauss–Jordan exploited the fact that the row system (7) was equivalent to (3) and (5). Is the column system (8) equivalent to (4) and (6)? The answer is 'Yes', and the reason for this will now be explained.

Suppose we perform a PS operation on Σ with α_{rs} as pivot entry. As we know, the row system of the new scheme Σ^* is the system obtained from (7) by the following DAWG step:

> for each $i \neq r$, *subtract* $(\alpha_{is}/\alpha_{rs})$ times the rth equation of (7) from the ith equation: *then divide* the rth equation by α_{rs}.

On the other hand, the PS recipe implies that the column system of Σ^* may be obtained from (8) by the following transposed DAWG step:

> for each $j \neq s$, *subtract* $(\alpha_{rj}/\alpha_{rs})$ times the sth equation of (8) from the jth equation: *then divide* the sth equation by $(-\alpha_{rs})$.

Since this operation is reversible, PS operations yield equivalent column systems.

In the final scheme, the row labels are $1, \ldots, n$ and the column labels are $n + 1, \ldots, 2n$, though not necessarily in the natural order.

$\mathbf{A}^{-1}$ is obtained from the final scheme exactly as in Example 2. Notice that the row system of the final scheme is essentially (5), except that the equations may have been reordered: finding $\mathbf{A}^{-1}\mathbf{b}$ is the same as setting $\mathbf{y} = \mathbf{0}$ and solving for $\mathbf{x}$. Thus the jth component of $\mathbf{A}^{-1}\mathbf{b}$ is the entry in the left border column, and the row *labelled* j, in the final scheme.

Similarly, the column system of the final scheme is essentially (6) except that equations may have been reordered. Finding $(\mathbf{A}^{\mathrm{T}})^{-1}\mathbf{c}$ is the same as setting $\mathbf{q} = \mathbf{0}$ and solving for $\mathbf{p}$. Thus the ith component of $(\mathbf{A}^{\mathrm{T}})^{-1}\mathbf{c}$ is the entry in the top row, and the column *labelled* $n + i$, in the final scheme.

Example 3 As in Example 2, let

$$\mathbf{A} = \begin{pmatrix} 0 & 3 & 4 \\ 2 & 0 & 3 \\ -1 & 1 & 0 \end{pmatrix}.$$

We wish to find (from scratch) $\mathbf{A}^{-1}$, $\mathbf{A}^{-1}\mathbf{b}$ and $(\mathbf{A}^{\mathrm{T}})^{-1}\mathbf{c}$ where

$$\mathbf{b} = \begin{pmatrix} 8 \\ 5 \\ 1 \end{pmatrix}, \quad \mathbf{c} = \begin{pmatrix} 5 \\ -6 \\ 2 \end{pmatrix}$$

Our initial scheme is

		1	2	3
	0	−5	6	−2
4	8	0	3	4
5	5	2	0	3
6	1	−1	①	0

Notice the signs in the top row! Pivoting as in Example 2 yields the following schemes:

		1	6	3
	−6	1	−6	−2
4	5	3	−3	4
5	5	②	0	3
2	1	−1	1	0

		5	6	3
	$-\frac{17}{2}$	$-\frac{1}{2}$	-6	$-\frac{7}{2}$
4	$-\frac{5}{2}$	$-\frac{3}{2}$	-3	$\boxed{-\frac{1}{2}}$
1	$\frac{5}{2}$	$\frac{1}{2}$	0	$\frac{3}{2}$
2	$\frac{7}{2}$	$\frac{1}{2}$	1	$\frac{3}{2}$

		5	6	4
	9	10	15	-7
3	5	3	6	-2
1	-5	-4	-9	3
2	-4	-4	-8	3

The inverse matrix $\mathbf{A}^{-1}$ has already been computed in Example 2 and we do not write it again. The components of $\mathbf{x} = \mathbf{A}^{-1}\mathbf{b}$ are read off from the left column (paying attention to labels) as

$$x_1 = -5, \quad x_2 = -4, \quad x_3 = 5.$$

Similarly, the components of $\mathbf{p} = (\mathbf{A}^{\mathrm{T}})^{-1}\mathbf{c}$ are read from the top row:

$$p_1 = -7, \quad p_2 = 10, \quad p_3 = 15.$$

Thus $\mathbf{A}^{-1}\mathbf{b} = \begin{pmatrix} -5 \\ -4 \\ 5 \end{pmatrix}$ and $(\mathbf{A}^{\mathrm{T}})^{-1}\mathbf{c} = \begin{pmatrix} -7 \\ 10 \\ 15 \end{pmatrix}$.

The sign convention for the column system (4) is not nearly as bizarre as might at first appear. Recall that the instructions for the pivot row r and pivot column s in a PS operation are

$$\beta_{rj} = \alpha_{rj}/\alpha_{rs}, \quad \text{for } j \neq s,$$

BUT

$$\beta_{is} = -\alpha_{is}/\alpha_{rs}, \quad \text{for } i \neq r.$$

This implies that a correct formulation of the column system must have some minus signs in it. The particular convention we have chosen involves a sign reversal at the beginning (enter $-\mathbf{c}^{\mathrm{T}}$ at the top of the first scheme) but not at the end (the components of $(\mathbf{A}^{\mathrm{T}})^{-1}\mathbf{c}$ are read off directly from the top row of the last scheme). This seems a logical way to proceed.

Nothing has been said up to now about the significance of the top-left-corner entry α_{00} in each scheme. Observe that

$$\text{if } \mathbf{Ax} = \mathbf{b} \quad \text{and} \quad \mathbf{A}^{\mathrm{T}}\mathbf{p} = \mathbf{c} \quad \text{then} \quad \mathbf{c}^{\mathrm{T}}\mathbf{x} = \mathbf{b}^{\mathrm{T}}\mathbf{p} = \mathbf{c}^{\mathrm{T}}\mathbf{A}^{-1}\mathbf{b}.$$

In Example 3 the common value of $\mathbf{c}^{\mathrm{T}}\mathbf{x}$ and $\mathbf{b}^{\mathrm{T}}\mathbf{p}$ is 9. The top-left-corner entry of the final scheme also happens to be 9. This is no accident, as we shall see in the next section.

Exercises

1. Repeat Exercise 3.1.1 using Tucker pivoting.
2. Given

$$\mathbf{A} = \begin{pmatrix} 1 & 0 & -1 & 4 \\ 2 & -1 & 1 & 0 \\ 1 & 0 & 0 & 3 \\ 2 & 1 & 3 & -1 \end{pmatrix}, \quad \mathbf{b} = \begin{pmatrix} 3 \\ -1 \\ 2 \\ -4 \end{pmatrix}, \quad \mathbf{c} = \begin{pmatrix} 6 \\ 4 \\ 3 \\ 5 \end{pmatrix}$$

find $\mathbf{A}^{-1}$, $\mathbf{A}^{-1}\mathbf{b}$ and $(\mathbf{A}^{\mathrm{T}})^{-1}\mathbf{c}$.

3.3 Complementary solutions

The last section used Tucker pivoting as a method of matrix inversion; but the mechanics of pivot-and-switch can be applied to schemes which are not square. The proximate reason for doing PS operations on rectangular schemes is that one obtains 'complementary solutions' to a pair of systems: the purpose of this section is to explain that procedure. The motivation for finding such complementary solutions lies in linear programming: this will become clear at the end of Chapter 5.

We say that n-vectors $\mathbf{x}$ and $\mathbf{q}$ are *complementary* if, for each $j = 1, \ldots, n$, at least one of the numbers x_j and q_j is zero. In this case we write $\mathbf{x}//\mathbf{q}$. Thus

$$\mathbf{x}//\mathbf{q} \text{ if and only if } x_j q_j = 0 \text{ for all } j.$$

For example, $(3 \quad 0 \quad 0 \quad 0 \quad 1)^{\mathrm{T}}//(0 \quad 2 \quad 0 \quad 4 \quad 0)^{\mathrm{T}}$.

Let $\mathbf{A}$ be an $m \times n$ matrix, $\mathbf{b}$ an m-vector, $\mathbf{c}$ an n-vector. Consider the pair of equation systems

$$\mathbf{Ax} + \mathbf{y} = \mathbf{b} \tag{1}$$

$$\mathbf{A}^{\mathrm{T}}\mathbf{p} - \mathbf{q} = \mathbf{c} \tag{2}$$

The system (1) describes m equations in the $n+m$ variables $x_1, \ldots, x_n, y_1, \ldots, y_m$. The system (2) describes n equations in the $m+n$ variables $p_1, \ldots, p_m, q_1, \ldots, q_n$. We say that the pair of $(n+m)$-vectors $(\mathbf{x} : \mathbf{y})$ and $(\mathbf{q} : \mathbf{p})$ form a *complementary solution* to (1) and (2) if they satisfy these equations and are complementary.

One complementary solution is immediate:

$$(\mathbf{x} : \mathbf{y}) = (\mathbf{0} : \mathbf{b}) \text{ and } (\mathbf{q} : \mathbf{p}) = (-\mathbf{c} : \mathbf{0}).$$

Let us see if we can find some others.

Keeping (1) as it is, and multiplying (2) by -1, we have the row and column systems for the rectangular scheme

$$\begin{array}{c|c|c}
 & & 1 \quad \cdots \quad n \\
\hline
 & 0 & -\mathbf{c}^{\mathrm{T}} \\
\hline
n+1 & & \\
\vdots & \mathbf{b} & \mathbf{A} \\
n+m & &
\end{array}$$

Suppose now that we perform some PS operations on this scheme, obtaining

$$\Sigma = \begin{array}{c|c|c}
 & & \pi_1 \quad \cdots \quad \pi_n \\
\hline
 & \lambda & -\tilde{\mathbf{c}}^{\mathrm{T}} \\
\hline
\pi_{n+1} & & \\
\vdots & \tilde{\mathbf{b}} & \tilde{\mathbf{A}} \\
\pi_{n+m} & &
\end{array}$$

where $(\pi_1, \ldots, \pi_{n+m})$ is some permutation of the integers $(1, \ldots, n+m)$.

Let $(\tilde{\mathbf{x}} : \tilde{\mathbf{y}})$ denote the vector $(\mathbf{x} : \mathbf{y})$ with its components rearranged in the order $\pi_1, \ldots, \pi_{n+m}$. Let $(\tilde{\mathbf{q}} : \tilde{\mathbf{p}})$ denote the same rearrangement of the components of $(\mathbf{q} : \mathbf{p})$. Then the row system of Σ is

$$\tilde{\mathbf{A}}\tilde{\mathbf{x}} + \tilde{\mathbf{y}} = \tilde{\mathbf{b}} \tag{3}$$

and the column system is

$$\tilde{\mathbf{A}}^{\mathrm{T}}\tilde{\mathbf{p}} - \tilde{\mathbf{q}} = \tilde{\mathbf{c}}. \tag{4}$$

We know from the last section, and the fact that moving from square to rectangular schemes changes nothing essential, that (3) is equivalent to (1) and (4) to (2). Thus another pair of solutions to (1) and (2) is given implicitly by

$$(\tilde{\mathbf{x}}:\tilde{\mathbf{y}})=(\mathbf{0}:\tilde{\mathbf{b}}),\quad (\tilde{\mathbf{q}}:\tilde{\mathbf{p}})=(-\tilde{\mathbf{c}}:\mathbf{0}). \tag{5}$$

Since $(\tilde{\mathbf{x}}:\tilde{\mathbf{y}})$ and $(\tilde{\mathbf{q}}:\tilde{\mathbf{p}})$ have been obtained from $(\mathbf{x}:\mathbf{y})$ and $(\mathbf{q}:\mathbf{p})$ by the same rearrangement of components, (5) gives a complementary solution.

The representation (5) of the complementary solution is *implicit*, being in terms of the reordered variables. To get an explicit representation we write Σ out in full as

$$\begin{array}{c|c|ccc}
 & & \pi_1 & \cdots & \pi_n \\
\hline
 & \alpha_{00} & \alpha_{01} & \cdots & \alpha_{0n} \\
\hline
\pi_{n+1} & \alpha_{10} & \alpha_{11} & \cdots & \alpha_{1n} \\
\vdots & \vdots & & \vdots & \\
\pi_{n+m} & \alpha_{m0} & \alpha_{m1} & \cdots & \alpha_{mn}
\end{array}$$

Then the complementary solution is read off as follows (here 'big' means greater than n, and 'small' means less than or equal to n):

Non-zero components of $\mathbf{x}$ correspond to small row labels:
if h $(\leq n)$ labels row i then $x_h=\alpha_{i0}$;
if h $(\leq n)$ labels a column then $x_h=0$.

Non-zero components of $\mathbf{y}$ correspond to big row labels:
if $n+l$ labels row i then $y_l=\alpha_{i0}$;
if $n+l$ labels a column then $y_l=0$.

Non-zero components of $\mathbf{q}$ correspond to small column labels:
if h $(\leq n)$ labels column j then $q_h=\alpha_{0j}$;
if h $(\leq n)$ labels a row then $q_h=0$.

Non-zero components of $\mathbf{p}$ correspond to big column labels:
if $n+l$ labels column j then $p_l=\alpha_{0j}$;
if $n+l$ labels a row then $p_l=0$.

Example Let

$$\mathbf{A}=\begin{pmatrix}3 & 4 & 5 & 6\\ 1 & -1 & 0 & 5\end{pmatrix},\quad \mathbf{b}=\begin{pmatrix}2\\ -1\end{pmatrix},\quad \mathbf{c}=\begin{pmatrix}6\\ -3\\ -5\\ 7\end{pmatrix}.$$

Find four different complementary solutions to the pair of systems

$$\mathbf{A}\mathbf{x} + \mathbf{y} = \mathbf{b}, \quad \mathbf{A}^{\mathrm{T}}\mathbf{p} - \mathbf{q} = \mathbf{c}.$$

We start with the scheme

		1	2	3	4
	0	-6	3	5	-7
5	2	3	4	$\textcircled{5}$	6
6	-1	1	-1	0	5

and the associated complementary solution

$$\text{(I)} \quad \mathbf{x} = \mathbf{0}, \quad \mathbf{y} = \mathbf{b}, \quad \mathbf{q} = -\mathbf{c}, \quad \mathbf{p} = \mathbf{0}.$$

Pivoting as shown we have the second scheme

		1	2	5	4
	-2	-9	-1	-1	-13
3	$\frac{2}{5}$	$\frac{3}{5}$	$\frac{4}{5}$	$\frac{1}{5}$	$\frac{6}{5}$
6	-1	$\textcircled{1}$	-1	0	5

Here the row labels are 3, which is 'small', and 6 $(=4+2)$, which is 'big'. Thus in the second complementary solution, $x_3 = \frac{2}{5}$, $y_2 = -1$ and all other components of $\mathbf{x}$ and $\mathbf{y}$ are zero. The column labels are 1, 2, 4 which are all 'small' and 5 $(=4+1)$ which is 'big': thus $q_1 = -9$, $q_2 = -1$, $q_4 = -13$, $p_1 = -1$ and all other components of $\mathbf{q}$ and $\mathbf{p}$ are zero. The second complementary solution is

$$\text{(II)} \quad \mathbf{x} = \begin{pmatrix} 0 \\ 0 \\ \frac{2}{5} \\ 0 \end{pmatrix}, \quad \mathbf{y} = \begin{pmatrix} 0 \\ -1 \end{pmatrix}, \quad \mathbf{q} = \begin{pmatrix} -9 \\ -1 \\ 0 \\ -13 \end{pmatrix}, \quad \mathbf{p} = \begin{pmatrix} -1 \\ 0 \end{pmatrix}.$$

Pivoting as shown, we have the third scheme

		6	2	5	4
	-11	9	-10	-1	32
3	1	$\textcircled{-\frac{3}{5}}$	$\frac{7}{5}$	$\frac{1}{5}$	$-\frac{9}{5}$
1	-1	1	-1	0	5

Here the row labels are 3 and 1: so in the third solution, $x_3 = 1$, $x_1 = -1$ and all other components of $(\mathbf{x} : \mathbf{y})$ are zero. The column labels are 6, 2, 5, 4: thus $p_2 = 9$, $q_2 = -10$, $p_1 = -1$, $q_4 = 32$ and all other components of $(\mathbf{q} : \mathbf{p})$ are zero. Thus the third complementary solution is

$$\text{(III)} \quad \mathbf{x} = \begin{pmatrix} -1 \\ 0 \\ 1 \\ 0 \end{pmatrix}, \quad \mathbf{y} = \begin{pmatrix} 0 \\ 0 \end{pmatrix}, \quad \mathbf{q} = \begin{pmatrix} 0 \\ -10 \\ 0 \\ 32 \end{pmatrix}, \quad \mathbf{p} = \begin{pmatrix} -1 \\ 9 \end{pmatrix}.$$

Pivoting as shown yields a fourth scheme:

		3	2	5	4
	4	15	11	2	5
6	$-\frac{5}{3}$	$-\frac{5}{3}$	$-\frac{7}{3}$	$-\frac{1}{3}$	3
1	$\frac{2}{3}$	$\frac{5}{3}$	$\frac{4}{3}$	$\frac{1}{3}$	2

and a fourth complementary solution

$$\text{(IV)} \quad \mathbf{x} = \begin{pmatrix} \frac{2}{3} \\ 0 \\ 0 \\ 0 \end{pmatrix}, \quad \mathbf{y} = \begin{pmatrix} 0 \\ -\frac{5}{3} \end{pmatrix}, \quad \mathbf{q} = \begin{pmatrix} 0 \\ 11 \\ 15 \\ 5 \end{pmatrix}, \quad \mathbf{p} = \begin{pmatrix} 2 \\ 0 \end{pmatrix}.$$

We must now enquire into the meaning of the top-left-corner entry of the scheme Σ, variously known as λ and α_{00}. Suppose we bring the top row into the row system of the *initial* scheme (with natural labelling). This means introducing a new variable y_0 and a new equation

$$-\mathbf{c}^{\mathrm{T}}\mathbf{x} + y_0 = 0.$$

A similar manoeuvre for the (equivalent) row system of Σ yields

$$-\tilde{\mathbf{c}}^{\mathrm{T}}\tilde{\mathbf{x}} + y_0 = \lambda.$$

Eliminating y_0,

$$\lambda = \mathbf{c}^{\mathrm{T}}\mathbf{x} - \tilde{\mathbf{c}}^{\mathrm{T}}\tilde{\mathbf{x}}. \tag{6}$$

Similarly, suppose we bring the left column into the equivalent column systems of the initial scheme and Σ. The relevant equations, involving a new variable q_0, are

$$q_0 - \mathbf{b}^{\mathrm{T}}\mathbf{p} = 0,$$
$$q_0 - \tilde{\mathbf{b}}^{\mathrm{T}}\tilde{\mathbf{p}} = \lambda.$$

Eliminating q_0,

$$\lambda = \mathbf{b}^T\mathbf{p} - \tilde{\mathbf{b}}^T\tilde{\mathbf{p}}. \tag{7}$$

Now let $(\mathbf{x}_S : \mathbf{y}_S)$ and $(\mathbf{q}_S : \mathbf{p}_S)$ denote the complementary solution corresponding to the scheme Σ. Recall from (5) that this is chosen such that $\tilde{\mathbf{x}}_S = \mathbf{0}$ and $\tilde{\mathbf{p}}_S = \mathbf{0}$. Thus from (6) and (7),

$$\lambda = \mathbf{c}^T\mathbf{x}_S = \mathbf{b}^T\mathbf{p}_S.$$

Hence

> the top-left entry of any scheme is the common value of $\mathbf{c}^T\mathbf{x}$ and $\mathbf{b}^T\mathbf{p}$ in the complementary solution read off from that scheme.

As an illustration of this, consider the third scheme and the solution (III) in the Example. The top left entry is -11. In fact,

$$\begin{aligned}\mathbf{c}^T\mathbf{x} &= 6 \times (-1) + (-5) \times 1 = -11 \\ &= 2 \times (-1) + (-1) \times 9 = \mathbf{b}^T\mathbf{p}.\end{aligned}$$

Exercise

Find four (or more) different complementary solutions to the pair of systems

$$\mathbf{A}\mathbf{x} + \mathbf{y} = \mathbf{b}, \quad \mathbf{A}^T\mathbf{p} - \mathbf{q} = \mathbf{c},$$

where

$$\mathbf{A} = \begin{pmatrix} 3 & 2 & 1 & -1 \\ 2 & 1 & -1 & 0 \\ 4 & -1 & 0 & 5 \end{pmatrix}, \quad \mathbf{b} = \begin{pmatrix} 1 \\ 0 \\ -1 \end{pmatrix}, \quad \mathbf{c} = \begin{pmatrix} 1 \\ 2 \\ 3 \\ 4 \end{pmatrix}.$$

For each solution, evaluate $\mathbf{c}^T\mathbf{x} = \mathbf{b}^T\mathbf{p}$.

4 Projections

This chapter is concerned with those aspects of linear algebra that are fundamental to econometrics: the theory of projections, the Choleski factorisation of positive definite matrices and the Gram-Schmidt orthogonalisation procedure. The material of this chapter is not used elsewhere in this book.

At a more abstract level, we are concerned with generalising the following two constructions in the plane.

(I) Let O be a given point and let L be a line through O. We can draw exactly one line, say M, which passes through O and is at right angles to L. Then L is the only line through O which is at right angles to M.

(II) Let L be a line and A a point not on L. Then we can find the closest point to A on L by dropping the perpendicular from A to L.

The really interesting construction is (II): its generalisation to n dimensions yields the method of least-squares estimation, the most popular procedure in econometrics.

Section 4.1 is mostly concerned with (I), but we do come to (II) at the end. Section 4.2 introduces symmetric matrices and discusses some of their properties. The material of the first two sections is brought together in Section 4.3, where we introduce the main concept of this chapter (projection matrices) and apply them to least-squares estimation. Section 4.4 deals with a method of constructing projections and least-squares estimates known as orthogonalisation: this in turn leads to the concept of pseudoinverse which is the nearest thing to an inverse that an arbitrary $m \times n$ matrix can have.

4.1 Orthogonal complements

In Section 1.4 we defined n-vectors $\mathbf{x}$ and $\mathbf{y}$ to be *orthogonal* if $\mathbf{x}^T\mathbf{y} = 0$. Let L be a linear subspace of R^n and define $L^\perp$ (read: 'L-orth') to be the set of all n-vectors which are orthogonal to *every* member of L. If $\mathbf{y}$ and $\mathbf{z}$ are in $L^\perp$ and α and β are scalars then

$$\mathbf{x}^T(\alpha\mathbf{y} + \beta\mathbf{z}) = \alpha\mathbf{x}^T\mathbf{y} + \beta\mathbf{x}^T\mathbf{z} = 0$$

for every member $\mathbf{x}$ of L. This means that $L^\perp$ is a linear subspace of R^n.

The subspace $L^\perp$ is called the *orthogonal complement* of L. The relationship between this kind of complement and the 'complementary' of Section 3.3 will be explained later in this section. In Section 9.1 we shall use the term 'complement' in a totally different sense.

If $n = 2$ and L has dimension 1, then L is a line through the origin. In this case, $L^\perp$ is the line through the origin at right angles to L, so we are indeed dealing with construction (I). Notice that in this two-dimensional case it is geometrically obvious that $(L^\perp)^\perp = L$. It will be shown in Theorem 2 that this holds in general.

Before doing this, we consider the orthogonal complements of ranges and kernels.

THEOREM 1 For any matrix $\mathbf{A}$,

$$(\text{rng } \mathbf{A})^\perp = \ker \mathbf{A}^T \quad \text{and} \quad (\ker \mathbf{A})^\perp = \text{rng } \mathbf{A}^T.$$

PROOF Let $\mathbf{A}$ be $m \times n$ and let $L = \text{rng } \mathbf{A}$, $M = \ker \mathbf{A}^T$; notice that these are linear subspaces of R^m. We begin by establishing the following results for m-vectors $\mathbf{b}$, $\mathbf{p}$.

(i) if $\mathbf{b} \in L$ and $\mathbf{p} \in M$ then $\mathbf{b}^T\mathbf{p} = 0$.
(ii) If $\mathbf{p}$ is not in M then $\mathbf{b}^T\mathbf{p} \neq 0$ for some $\mathbf{b}$ in L.
(iii) If $\mathbf{b}$ is not in L then $\mathbf{b}^T\mathbf{p} \neq 0$ for some $\mathbf{p}$ in M.

If $\mathbf{p}$ is in M then $(\mathbf{Ax})^T\mathbf{p} = \mathbf{x}^T\mathbf{A}^T\mathbf{p} = 0$ for each $\mathbf{x}$ in R^n: this proves (i). If $\mathbf{p}$ is not in M we can choose some j such that component j of $\mathbf{A}^T\mathbf{p}$ is non-zero; letting $\mathbf{b}$ be column j of $\mathbf{A}$ we have $\mathbf{b} \in L$ and $\mathbf{b}^T\mathbf{p} \neq 0$: this proves (ii). We said in Section 2.4 that the Theorem of the Alternative was going to be useful in discussing orthogonality and now we see why: it is (iii).

By (i) and (ii), $L^\perp = M$: thus $(\text{rng } \mathbf{A})^\perp = \ker \mathbf{A}^T$.

By (i) and (iii), $M^\perp = L$: thus $(\ker \mathbf{A}^T)^\perp = \text{rng } \mathbf{A}$. Applying this with $\mathbf{A}$ replaced by $\mathbf{A}^T$, and using $(\mathbf{A}^T)^T = \mathbf{A}$, we see that $(\ker \mathbf{A})^\perp = \text{rng } \mathbf{A}^T$. QED

The relevance of Theorem 1 to the general theory of orthogonal complements lies in the fact that any linear subspace L of R^n can be expressed as the range of a matrix $\mathbf{X}$, with n rows. To see this, let the dimension of L be k: if $k = 0$, the zero n-vector will do; if $k > 0$, let $\mathbf{X}$ be a basis for L.

THEOREM 2 For any linear subspace L of R^n,

$$\dim L + \dim L^{\perp} = n \quad \text{and} \quad (L^{\perp})^{\perp} = L.$$

PROOF Let $\mathbf{X}$ be a matrix whose range is L. By Theorem 1, $L^{\perp} = \ker \mathbf{X}^{\mathrm{T}}$. Since $\mathbf{X}$ has n *rows*, we infer from Theorem 2.4.1 that

$$\dim L^{\perp} = \text{nul } \mathbf{X}^{\mathrm{T}} = n - \text{rng } \mathbf{X} = n - \dim L.$$

Also, by Theorem 1,

$$(L^{\perp})^{\perp} = \text{rng } (\mathbf{X}^{\mathrm{T}})^{\mathrm{T}} = \text{rng } \mathbf{X} = L. \qquad \text{QED}$$

We are now in a position to prove the general property of linear subspaces mentioned in Section 2.3: any linear subspace can be expressed both as the range of a matrix and as the kernel of another matrix.

THEOREM 3 For any linear subspace L of R^n there exist matrices $\mathbf{X}$ and $\mathbf{A}$ such that

$$L = \text{rng } \mathbf{X} = \ker \mathbf{A}.$$

PROOF We already know about $\mathbf{X}$. Similarly, there exists a matrix $\mathbf{Z}$ such that $L^{\perp} = \text{rng } \mathbf{Z}$. By Theorems 1 and 2,

$$\ker \mathbf{Z}^{\mathrm{T}} = (L^{\perp})^{\perp} = L.$$

Setting $\mathbf{A} = \mathbf{Z}^{\mathrm{T}}$ we have $L = \ker \mathbf{A}$. QED

We now state some simple facts about orthogonal complements. First, it is obvious that

$$\{\mathbf{0}\}^{\perp} = R^n \quad \text{and} \quad (R^n)^{\perp} = \{\mathbf{0}\}. \tag{1}$$

Second, for any linear subspace L

$$L \cap L^{\perp} = \{\mathbf{0}\}. \tag{2}$$

The reason for this is that if $\mathbf{x}$ belongs both to L and to $L^{\perp}$ then $\mathbf{x}^{\mathrm{T}}\mathbf{x} = 0$, so $\mathbf{x} = \mathbf{0}$ by Theorem 1.4.3.

Third, let L be a k-dimensional linear subspace of R^n, and suppose that $0 < k < n$. Let $\mathbf{X}$ and $\mathbf{Z}$ be bases for L and $L^\perp$ respectively. Then $\mathbf{X}$ is an $n \times k$ matrix; and by Theorem 2, $\mathbf{Z}$ is $n \times (n-k)$. Let the k-vector $\mathbf{b}$ and the $(n-k)$-vector $\mathbf{c}$ be such that

$$\mathbf{Xb} + \mathbf{Zc} = \mathbf{0}.$$

Then by (2), $\mathbf{Xb} = \mathbf{0} = \mathbf{Zc}$: and since $\mathbf{X}$ and $\mathbf{Z}$ are bases, this implies that $\mathbf{b} = \mathbf{0}$ and $\mathbf{c} = \mathbf{0}$. We have now shown that the $n \times n$ matrix

$$\mathbf{B} = (\mathbf{X}\ \mathbf{Z})$$

is nonsingular.

Thus for any k-dimensional linear subspace L of R^n (with $0 < k < n$) there exists a basis $\mathbf{B}$ for R^n whose first k columns form a basis for L and whose remaining columns form a basis for $L^\perp$. This establishes the relationship between orthogonal complements and complementarity. Given any $\mathbf{x}$ in L and any $\mathbf{z}$ in $L^\perp$ we may write

$$\mathbf{x} = \mathbf{Bf} \quad \text{and} \quad \mathbf{z} = \mathbf{Bg}$$

where the first k components of $\mathbf{g}$ and the last $n-k$ components of $\mathbf{f}$ are all zcro: thus $\mathbf{B}^{-1}\mathbf{x} // \mathbf{B}^{-1}\mathbf{z}$.

Since $\mathbf{B}$ is a basis for R^n, any n-vector $\mathbf{y}$ can be expressed as a linear combination of the columns of $\mathbf{B}$. Grouping the first k terms in this linear combination, and the last $n-k$ terms, we express $\mathbf{y}$ as the sum of a member of L and a member of $L^\perp$. This turns out to be construction (II) in n-space, as the next theorem demonstrates.

THEOREM 4 Let L be a linear subspace of R^n, $\mathbf{y}$ an n-vector. Then there exists a vector $\mathbf{y}_L$, uniquely determined by $\mathbf{y}$ and L, such that

$$\mathbf{y}_L \in L \quad \text{and} \quad \mathbf{y} - \mathbf{y}_L \in L^\perp \tag{3}$$

The vector $\mathbf{y}_L$ is the member of L that is closest to $\mathbf{y}$: if $\mathbf{x} \in L$ then

$$\|\mathbf{y} - \mathbf{x}\| \geq \|\mathbf{y} - \mathbf{y}_L\|$$

with equality only if $\mathbf{x} = \mathbf{y}_L$.

PROOF We show first that there exists some $\mathbf{y}_L$ satisfying (3). Let $\dim L = k$. We know that if $0 < k < n$ then $\mathbf{y}$ can be written in the form $\mathbf{y}_L + \mathbf{y}_M$ where $\mathbf{y}_L \in L$ and $\mathbf{y}_M \in L^\perp$. If k is either 0 or n we can apply (1): $\mathbf{y}_L - \mathbf{0}$ if $k = 0$, $\mathbf{y}_L = \mathbf{y}$ if $k = n$.

We now prove that $\mathbf{y}_L$ is uniquely determined by $\mathbf{y}$ and L. Let $\mathbf{x}_1$ and $\mathbf{x}_2$ be members of L such that $\mathbf{y} - \mathbf{x}_1$ and $\mathbf{y} - \mathbf{x}_2$ belong to $L^\perp$.

Clearly $\mathbf{x}_1 - \mathbf{x}_2 \in L$. Also

$$\mathbf{x}_1 - \mathbf{x}_2 = (\mathbf{y} - \mathbf{x}_2) - (\mathbf{y} - \mathbf{x}_1)$$

so $\mathbf{x}_1 - \mathbf{x}_2 \in L^\perp$. By (2), $\mathbf{x}_1 = \mathbf{x}_2$.

To prove the last part, let $\mathbf{x}$ be a member of L other than $\mathbf{y}_L$ and let

$$\alpha = \|\mathbf{x} - \mathbf{y}\|, \quad \beta = \|\mathbf{y}_L - \mathbf{y}\|, \quad \gamma = \|\mathbf{x} - \mathbf{y}_L\|.$$

Since $\mathbf{x} \neq \mathbf{y}_L$, $\gamma > 0$: we want to show that $\alpha > \beta$. Now

$$\mathbf{x} - \mathbf{y} = (\mathbf{x} - \mathbf{y}_L) + (\mathbf{y}_L - \mathbf{y}).$$

Also $\alpha^2 = (\mathbf{x} - \mathbf{y})^{\mathrm{T}}(\mathbf{x} - \mathbf{y})$ and similar expressions hold for β^2 and γ^2. Thus

$$\alpha^2 = \gamma^2 + 2\theta + \beta^2$$

where $\theta = (\mathbf{x} - \mathbf{y}_L)^{\mathrm{T}}(\mathbf{y}_L - \mathbf{y})$. But $\mathbf{x}$ and $\mathbf{y}_L$ are in L and $\mathbf{y} - \mathbf{y}_L \in L^\perp$. Thus $\theta = 0$ and $\alpha = \sqrt{\beta^2 + \gamma^2} > \beta$. QED

The vector $\mathbf{y}_L$ in Theorem 4 is called the *projection* of $\mathbf{y}$ on L. We proved its existence by expressing $\mathbf{y}$ as a linear combination of the columns of a nonsingular matrix β whose first k columns span L and whose other columns span $L^\perp$. This procedure may be awkward to implement in practice and it is natural to ask whether $\mathbf{y}_L$ can be constructed from $\mathbf{y}$ and L without constructing bases both for L and for $L^\perp$. The way this is done will be explained in Section 4.3 after we have established some results on symmetric matrices.

Exercises

1. Let L be the range of the matrix

$$\begin{pmatrix} 1 & 2 \\ 1 & 3 \\ 1 & 4 \\ 1 & 5 \end{pmatrix}$$

Find a basis for $L^\perp$.

2. Prove that for any linear subspaces K and L of R^n

$$(K + L)^\perp = K^\perp \cap L^\perp \quad \text{and} \quad (K \cap L)^\perp = K^\perp + L^\perp$$

where $+$ is defined as in Exercise 2.3.3.

4.2 Symmetric matrices

A square matrix $\mathbf{S}$ is said to be *symmetric* if $\mathbf{S}^{\mathrm{T}} = \mathbf{S}$: thus

$$\begin{pmatrix} 1 & 2 & 0 \\ 2 & 3 & 4 \\ 0 & 4 & 5 \end{pmatrix}$$

is symmetric and so is any diagonal matrix.

Three important properties of symmetric matrices follow almost immediately from the definition.

(I) If $\mathbf{S}$ is a symmetric matrix of order n, and $\mathbf{A}$ is a matrix with n columns, then

$$(\mathbf{ASA}^{\mathrm{T}})^{\mathrm{T}} = \mathbf{A}^{\mathrm{TT}}\mathbf{S}^{\mathrm{T}}\mathbf{A}^{\mathrm{T}} = \mathbf{ASA}^{\mathrm{T}}$$

so the matrix $\mathbf{ASA}^{\mathrm{T}}$ is symmetric. Since identity matrices are symmetric, it follows that $\mathbf{AA}^{\mathrm{T}}$ and $\mathbf{A}^{\mathrm{T}}\mathbf{A}$ are symmetric for every $\mathbf{A}$.

(II) Given a square matrix $\mathbf{B}$ we can define the matrices

$$\mathbf{B}^2 = \mathbf{BB}, \quad \mathbf{B}^3 = \mathbf{B}^2\mathbf{B}, \quad \text{and so on.}$$

If $\mathbf{S}$ is symmetric, then $\mathbf{S}^k$ is symmetric for every k: for in this case $\mathbf{S}^2 = \mathbf{S}^{\mathrm{T}}\mathbf{S}$ and

$$\mathbf{S}^k = \mathbf{S}^{\mathrm{T}}\mathbf{S}^{k-2}\mathbf{S} \quad \text{for } k > 2,$$

so the required result follows from (I).

(III) If $\mathbf{S}$ is symmetric and nonsingular then

$$(\mathbf{S}^{-1})^{\mathrm{T}} = (\mathbf{S}^{\mathrm{T}})^{-1} = \mathbf{S}^{-1}$$

so $\mathbf{S}^{-1}$ is also symmetric.

Symmetric matrices arise in a wide variety of contexts in mathematics, notably in connection with quadratic forms. A *quadratic form* in n variables $x_1, \ldots, x_n$ is an expression of the form

$$f(\mathbf{x}) = \sum_{i=1}^{n} \sum_{j=1}^{n} a_{ij}x_i x_j = \mathbf{x}^{\mathrm{T}}\mathbf{Ax} \tag{1}$$

where $\mathbf{A}$ is a given square matrix of order n. For example if

$$\mathbf{A} = \begin{pmatrix} 1 & 2 \\ 3 & 4 \end{pmatrix}$$

then $f(\mathbf{x}) = x_1^2 + 2x_1 x_2 + 3x_2 x_1 + 4x_2^2$. Since $x_1 x_2 = x_2 x_1$ we may write

$$f(\mathbf{x}) = x_1^2 + 5x_1 x_2 + 4x_2^2.$$

Clearly, a similar trick can be played with any quadratic form; the coefficient of $x_i x_j$ in (1) is

$$a_{ii} \quad \text{if } i = j, \quad a_{ij} + a_{ji} \quad \text{if } i \neq j.$$

It follows that $\mathbf{x}^T\mathbf{A}\mathbf{x} = \mathbf{x}^T\mathbf{S}\mathbf{x}$, where $\mathbf{S}$ is the *symmetric* matrix $\frac{1}{2}\mathbf{A} + \frac{1}{2}\mathbf{A}^T$. In other words, every quadratic form has a symmetric matrix associated with it, and vice versa.

We say that a symmetric matrix $\mathbf{S}$ is *positive semidefinite* if $\mathbf{x}^T\mathbf{S}\mathbf{x} \geq 0$ for every $\mathbf{x}$. The previous paragraph suggests a natural way of defining positive semidefiniteness for an arbitrary square matrix $\mathbf{A}$: $\mathbf{A}$ is positive semidefinite if $\mathbf{A} + \mathbf{A}^T$ is. But for the rest of this book, the term 'positive semidefinite' (and, later, 'positive definite') will be applied *only* to symmetric matrices: and when we say that a matrix is positive semidefinite, symmetry will be understood.

THEOREM 1 Let $\mathbf{S}$ be positive semidefinite. Then $\mathbf{x}^T\mathbf{S}\mathbf{x} = 0$ if and only if $\mathbf{S}\mathbf{x} = \mathbf{0}$.

PROOF 'If' is obvious. To prove 'only if' suppose that $\mathbf{S}\mathbf{x} \neq \mathbf{0}$ and let $\mathbf{y} = \mathbf{S}\mathbf{x}$. We wish to show that $\mathbf{x}^T\mathbf{y} > 0$.

Since $\mathbf{S}$ is positive semidefinite,

$$(\mathbf{x} - \alpha\mathbf{y})^T\mathbf{S}(\mathrm{x} - \alpha\mathbf{y}) \geq 0 \tag{2}$$

for any scalar α. Since $\mathbf{S}$ is symmetric, $\mathbf{x}^T\mathbf{S}\mathbf{y} = (\mathbf{S}\mathbf{x})^T\mathbf{y}$. Hence the left-hand side of (2) may be written

$$\mathbf{x}^T\mathbf{y} - 2\alpha\mathbf{y}^T\mathbf{y} + \alpha^2\mathbf{y}^T\mathbf{S}\mathbf{y}$$

It follows that

$$\mathbf{x}^T\mathbf{y} \geq \alpha(2\mathbf{y}^T\mathbf{y} - \alpha\mathbf{y}^T\mathbf{S}\mathbf{y}) \tag{3}$$

for every scalar α. Since $\mathbf{y} \neq \mathbf{0}$, $\mathbf{y}^T\mathbf{y} > 0$; thus the right-hand side of (3) is positive for all sufficiently small positive α. Hence $\mathbf{x}^T\mathbf{y} > 0$.

QED

We define a symmetric matrix $\mathbf{S}$ to be *positive definite* if it is positive semidefinite and nonsingular. It follows immediately from Theorem 1 that a symmetric matrix $\mathbf{S}$ is positive definite if and only if

$$\mathbf{x}^T\mathbf{S}\mathbf{x} > 0 \quad \text{for every non-zero } \mathbf{x}.$$

All the diagonal entries of a positive definite matrix are positive (let $\mathbf{x}$ be a column of $\mathbf{I}$). On the other hand, positivity of all diagonal entries is not a *sufficient* condition for a symmetric matrix to be positive definite: for if

$$\mathbf{S} = \begin{pmatrix} 1 & 2 \\ 2 & 1 \end{pmatrix}$$

then $(1, -1)\mathbf{S}(1, -1)^{\mathrm{T}} = -2$, so $\mathbf{S}$ is not even positive semidefinite.

We can bring these concepts right into the mainstream of linear algebra by noting that *any* $m \times n$ matrix $\mathbf{A}$ has two positive semidefinite matrices associated with it. These are the $m \times m$ matrix $\mathbf{AA}^{\mathrm{T}}$ and the $n \times n$ matrix $\mathbf{A}^{\mathrm{T}}\mathbf{A}$. These matrices are symmetric by (I) above: also

$$\mathbf{x}^{\mathrm{T}}(\mathbf{A}^{\mathrm{T}}\mathbf{A})\mathbf{x} = (\mathbf{Ax})^{\mathrm{T}}(\mathbf{Ax}) \geq 0$$

for any n-vector $\mathbf{x}$, and a similar argument holds for $\mathbf{AA}^{\mathrm{T}}$.

We now discuss the relationships between $\mathbf{A}$ and its associated positive semidefinite matrices.

THEOREM 2 For any matrix $\mathbf{A}$, rng $\mathbf{A}^{\mathrm{T}}\mathbf{A} =$ rng $\mathbf{A}^{\mathrm{T}}$.

PROOF Let $\mathbf{A}$ have n columns and let $L = \ker \mathbf{A}$. Let $\mathbf{x}$ be any n-vector. If $\mathbf{A}^{\mathrm{T}}\mathbf{Ax} = \mathbf{0}$ then

$$(\mathbf{Ax})^{\mathrm{T}}(\mathbf{Ax}) = \mathbf{x}^{\mathrm{T}}(\mathbf{A}^{\mathrm{T}}\mathbf{Ax}) = 0$$

so $\mathbf{Ax} = \mathbf{0}$. Conversely, if $\mathbf{Ax} = \mathbf{0}$ then $\mathbf{A}^{\mathrm{T}}\mathbf{Ax} = \mathbf{0}$. Thus $L = \ker \mathbf{A}^{\mathrm{T}}\mathbf{A}$. But then, by Theorem 4.1.1 and the symmetry of $\mathbf{A}^{\mathrm{T}}\mathbf{A}$,

$$\text{rng } \mathbf{A}^{\mathrm{T}}\mathbf{A} = L^{\perp} = \text{rng } \mathbf{A}^{\mathrm{T}}. \qquad \text{QED}$$

Applying Theorem 2 to $\mathbf{A}^{\mathrm{T}}$ we see that rng $\mathbf{AA}^{\mathrm{T}} =$ rng $\mathbf{A}$. Taking dimensions, rk $\mathbf{A}^{\mathrm{T}}\mathbf{A} =$ rk $\mathbf{A}^{\mathrm{T}}$ and rk $\mathbf{AA}^{\mathrm{T}} =$ rk $\mathbf{A}$. But we know from Theorem 2.4.1 that rk $\mathbf{A}^{\mathrm{T}} =$ rk $\mathbf{A}$. Putting all these equations together we have the following theorem.

THEOREM 3 (Rank Theorem)
Let $\mathbf{A}$ be any matrix. Then each of the matrices $\mathbf{A}^{\mathrm{T}}$, $\mathbf{A}^{\mathrm{T}}\mathbf{A}$ and $\mathbf{AA}^{\mathrm{T}}$ has the same rank as $\mathbf{A}$.

We have shown above that $\mathbf{A}^{\mathrm{T}}\mathbf{A}$ is positive semidefinite for every $\mathbf{A}$. The question now arises, when will $\mathbf{A}^{\mathrm{T}}\mathbf{A}$ be positive definite?

We know from our earliest discussion of rank, at the end of Section 3.3, that the following three statements are equivalent:
(α) the columns of $\mathbf{A}$ are linearly independent;
(β) the rank of $\mathbf{A}$ is its total number of columns;
(γ) $\mathbf{A}$ is a basis for its range.
When one and hence all of (α), (β) and (γ) hold we say that $\mathbf{A}$ has *full column rank*. By the Rank Theorem, $\mathbf{A}$ has full column rank if and only if the square matrix $\mathbf{A}^T\mathbf{A}$ is nonsingular. This gives us the following theorem.

THEOREM 4 $\mathbf{A}^T\mathbf{A}$ is positive definite if and only if $\mathbf{A}$ has full column rank.

The last theorem of this section implies that *any* positive definite matrix can be expressed in the form $\mathbf{A}^T\mathbf{A}$ where $\mathbf{A}$ has full column rank. For its proof and for Section 4.4, it is helpful to have a special notation for diagonal matrices: the $n \times n$ diagonal matrix with diagonal entries $\mu_1, \ldots, \mu_n$ will be denoted by

$$\text{diag}\,(\mu_1, \ldots, \mu_n).$$

THEOREM 5 (Choleski Factorisation Theorem)
If $\mathbf{S}$ is positive definite there exists an upper triangular matrix $\mathbf{A}$, with all diagonal entries positive, such that $\mathbf{S} = \mathbf{A}^T\mathbf{A}$.
PROOF Let $\mathbf{S}$ be a positive definite $n \times n$ matrix, let $\alpha = s_{11}$ and let

$$\mathbf{b} = (s_{21}/s_{11}, \ldots, s_{n1}/s_{11})^T.$$

One step of Gaussian elimination has the same effect on $\mathbf{S}$ as premultiplication by the square matrix $\mathbf{E}$ whose first column is $(1\colon -\mathbf{b})$ and whose other columns are as in $\mathbf{I}_n$. Since $\mathbf{S}$ is symmetric, the first row of $\mathbf{S}$ is $(\alpha, \mathbf{b}^T)$. Thus for some square matrix $\mathbf{W}$ of order $n-1$, we have

$$\mathbf{ES} = \begin{pmatrix} \alpha & \alpha\mathbf{b}^T \\ \mathbf{0} & \mathbf{W} \end{pmatrix} \quad \text{whence} \quad \mathbf{ESE}^T = \begin{pmatrix} \alpha & \mathbf{0} \\ \mathbf{0} & \mathbf{W} \end{pmatrix}.$$

Since $\mathbf{S}$ is positive definite and $\mathbf{E}$ nonsingular, $\mathbf{ESE}^T$ is positive definite. Thus the diagonal entries of $\mathbf{W}$ are positive and elimination can proceed without row interchange.

This procedure can be repeated until elimination is complete: no row interchange is ever required. We have

$$\mathbf{GS} = \mathbf{U},$$

where $\mathbf{G}$ is lower triangular with 1's down the diagonal and $\mathbf{U}$ is upper triangular with positive diagonal entries $\mu_1, \ldots, \mu_n$.

Since $\mathbf{S}$ is symmetric we may apply the same trick to $\mathbf{G}$ that we applied above to the single elimination step $\mathbf{E}$:

$$\mathbf{GSG}^{\mathrm{T}} = \mathbf{D} = \operatorname{diag}(\mu_1, \ldots, \mu_n).$$

But then $\mathbf{UG}^{\mathrm{T}} = \mathbf{D}$, so $(\mathbf{G}^{\mathrm{T}})^{-1} = \mathbf{D}^{-1}\mathbf{U}$. Transposing, and using the fact that $\mathbf{D}$ is symmetric, we have $\mathbf{G}^{-1} = \mathbf{U}^{\mathrm{T}}\mathbf{D}^{-1}$, whence

$$\mathbf{S} = \mathbf{G}^{-1}\mathbf{U} = \mathbf{U}^{\mathrm{T}}\mathbf{D}^{-1}\mathbf{U}.$$

Defining the square matrices $\mathbf{Q} = \operatorname{diag}(\sqrt{\mu_1}, \ldots, \sqrt{\mu_n})$, $\mathbf{A} = \mathbf{Q}^{-1}\mathbf{U}$, we have $\mathbf{S} = \mathbf{A}^{\mathrm{T}}\mathbf{A}$. Here $\mathbf{A}$ is upper triangular with positive diagonal entries $\sqrt{\mu_1}, \ldots, \sqrt{\mu_n}$. QED

The proof of Theorem 5 gives a quick way of inverting a positive definite matrix $\mathbf{S}$. Gaussian elimination on $(\mathbf{S}, \mathbf{I})$ yields the pair of triangular matrices $(\mathbf{U}, \mathbf{G})$. Let $\mathbf{D}$ be the diagonal matrix with the same diagonal entries as $\mathbf{U}$: we know that $\mathbf{S} = \mathbf{U}^{\mathrm{T}}\mathbf{D}^{-1}\mathbf{U}$ and $\mathbf{UG}^{\mathrm{T}} = \mathbf{D}$. Hence $\mathbf{S}^{-1}$ may be computed as $\mathbf{G}^{\mathrm{T}}\mathbf{D}^{-1}\mathbf{G}$. Square roots are not required.

Example Invert the positive definite matrix

$$\mathbf{S} = \begin{pmatrix} 1 & 2 & 1 \\ 2 & 7 & -1 \\ 1 & -1 & 10 \end{pmatrix}.$$

Starting with $(\mathbf{S}, \mathbf{I})$ and performing two elimination steps, we have

$$\left.\begin{array}{rrr} 1 & 2 & 1 \\ 0 & 3 & -3 \\ 0 & 0 & 6 \end{array}\right| \begin{array}{rrr} 1 & 0 & 0 \\ -2 & 1 & 0 \\ -3 & 1 & 1 \end{array}$$

Thus $\mathbf{D} = \operatorname{diag}(1, 3, 6)$ and

$$\mathbf{G}^{\mathrm{T}} = \begin{pmatrix} 1 & -2 & -3 \\ 0 & 1 & 1 \\ 0 & 0 & 1 \end{pmatrix}.$$

Recalling that $\mathbf{S}^{-1} = \mathbf{G}^{\mathrm{T}}\mathbf{D}^{-1}\mathbf{G}$ and using symmetry we have

$$\mathbf{S}^{-1} = \begin{pmatrix} \frac{1}{1}+\frac{4}{3}+\frac{9}{6} & \cdot & \cdot \\ -\frac{2}{3}-\frac{3}{6} & \frac{1}{3}+\frac{1}{6} & \cdot \\ -\frac{3}{6} & \frac{1}{6} & \frac{1}{6} \end{pmatrix}$$

$$= (\tfrac{1}{6})\begin{pmatrix} 23 & -7 & -3 \\ -7 & 3 & 1 \\ -3 & 1 & 1 \end{pmatrix}.$$

This example is a bit of a swindle: we were 'lucky' enough to start with a matrix which 'happened' to be positive definite. The reason why it is useful to have a special method for the inversion of positive definite matrices is that situations arise (especially in statistics) where one is required to invert a matrix which is positive definite *by construction*. The simplest example is: given the matrix $\mathbf{X}$ of full column rank, find $(\mathbf{X}^{\mathrm{T}}\mathbf{X})^{-1}$. We shall return to this point at the end of the next section.

Exercises

1. Prove that the inverse of a positive definite matrix is positive definite.

2. Invert the positive definite matrix

$$\begin{pmatrix} 4 & 1 & 2 \\ 1 & 4 & 5 \\ 2 & 5 & 7 \end{pmatrix}.$$

3. 'Choleski factorisation of the positive definite quadratic form $\mathbf{x}^{\mathrm{T}}\mathbf{S}\mathbf{x}$ yields the form $(\mathbf{A}\mathbf{x})^{\mathrm{T}}(\mathbf{A}\mathbf{x})$. This is nothing more than the school algebra trick of completing the square.' Explain, with examples.

4.3 Least squares

Our first object in this section is to complete the unfinished business left over from Section 4.1. Let L be a linear subspace of R^n, and for each $\mathbf{y}$ in R^n let $\mathbf{y}_L$ denote the projection of $\mathbf{y}$ on L. We want to find $\mathbf{y}_L$ in terms of $\mathbf{y}$ and L without constructing bases both for L and for $L^{\perp}$.

THEOREM 1 Given a linear subspace L of R^n there exists exactly one matrix $\mathbf{P}$ such that $\mathbf{Py} = \mathbf{y}_L$ for all $\mathbf{y}$. If dim $L = 0$ then $\mathbf{P} = \mathbf{O}$. If dim $L > 0$ and $\mathbf{X}$ is any basis for L then

$$\mathbf{P} = \mathbf{X}(\mathbf{X}^{\mathrm{T}}\mathbf{X})^{-1}\mathbf{X}^{\mathrm{T}}. \tag{1}$$

PROOF Since $\mathbf{y}_L$ is uniquely determined by $\mathbf{y}$ and L, there can be at most one matrix with the required property. If $L = \{\mathbf{0}\}$ then $\mathbf{O}$ does the trick. Now suppose that L has a basis $\mathbf{X}$. By Theorem 4.2.4, the matrix $\mathbf{X}^{\mathrm{T}}\mathbf{X}$ is positive definite, so we can define a matrix $\mathbf{P}$ by (1). Let $\mathbf{y}$ be any n-vector: we wish to show that $\mathbf{Py} = \mathbf{y}_L$.

Obviously $\mathbf{Py}$ is in the range of $\mathbf{X}$: also $\mathbf{X}^{\mathrm{T}}\mathbf{P} = \mathbf{X}^{\mathrm{T}}$ so $\mathbf{X}^{\mathrm{T}}(\mathbf{y} - \mathbf{Py}) = \mathbf{0}$. But since $\mathbf{X}$ is a basis for L, rng $\mathbf{X} = L$, whence ker $\mathbf{X}^{\mathrm{T}} = L^{\perp}$. Thus $\mathbf{Py} \in L$ and $\mathbf{y} - \mathbf{Py} \in L^{\perp}$, so $\mathbf{Py} = \mathbf{y}_L$. QED

The matrix $\mathbf{P}$ of Theorem 1 is called the *projection* onto L. We now ask the question, what kinds of matrices qualify as projections?

Let $\mathbf{P}$ be as in (1). Since $\mathbf{X}^{\mathrm{T}}\mathbf{X}$ is symmetric, so is the matrix $\mathbf{S} = (\mathbf{X}^{\mathrm{T}}\mathbf{X})^{-1}$. Hence $\mathbf{P} = \mathbf{XSX}^{\mathrm{T}}$ is also symmetric. Now $\mathbf{X}^{\mathrm{T}}\mathbf{XS} = \mathbf{I}_k$, where k is the number of columns of $\mathbf{X}(k = \dim L)$. Thus

$$\mathbf{P}^2 = (\mathbf{XS})(\mathbf{X}^{\mathrm{T}}\mathbf{XS})\mathbf{X}^{\mathrm{T}} = \mathbf{XSX}^{\mathrm{T}} = \mathbf{P}$$

Hence the projection $\mathbf{P}$ onto any subspace of positive dimension has the properties

$$\mathbf{P}^{\mathrm{T}} = \mathbf{P}, \tag{2}$$

$$\mathbf{P}^2 = \mathbf{P}. \tag{3}$$

Notice that $\mathbf{O}$ also has these properties, so the 'positive dimension' qualification may be dropped.

Property (3) has a name. A square matrix $\mathbf{G}$ is said to be *idempotent* if $\mathbf{G}^2 = \mathbf{G}$. For example

$$\begin{pmatrix} 1 & 0 \\ 0 & 0 \end{pmatrix} \quad \text{and} \quad \begin{pmatrix} \frac{2}{3} & \frac{1}{3} \\ \frac{2}{3} & \frac{1}{3} \end{pmatrix}$$

are idempotent matrices: the former is symmetric, the latter is not. if $\mathbf{G}$ is idempotent then

$$\mathbf{G}^3 = \mathbf{G}^2\mathbf{G} = \mathbf{GG} = \mathbf{G}, \quad \mathbf{G}^4 = \mathbf{G}^3\mathbf{G} = \mathbf{GG} = \mathbf{G},$$

and so on. Thus an idempotent matrix remains the same when raised to any power. (*Potens* is Latin for 'power', *idem* for 'same'.)

THEOREM 2 If $\mathbf{P}$ is the projection onto a linear subspace L of R^n then $\mathbf{P}$ is symmetric and idempotent, and rng $\mathbf{P} = L$.

Conversely, any symmetric idempotent matrix is the projection onto its range.

PROOF Let $\mathbf{P}$ be the projection onto L. We showed above that $\mathbf{P}$ is symmetric and idempotent. For any $\mathbf{x}$ in L the closest point to $\mathbf{x}$ in L is $\mathbf{x}$ itself, so $\mathbf{x} = \mathbf{Px}$. Thus $L \subset \operatorname{rng} \mathbf{P}$: but $\operatorname{rng} \mathbf{P} \subset L$ by definition of a projection, so $\operatorname{rng} \mathbf{P} = L$.

Conversely, let $\mathbf{P}$ be any symmetric idempotent matrix and let $L = \operatorname{rng} \mathbf{P}$. Then for any $\mathbf{y}$ we have $\mathbf{Py} \in L$ and it remains to show that $\mathbf{y} - \mathbf{Py} \in L^{\perp}$. But $L^{\perp} = \ker \mathbf{P}$ by the symmetry of $\mathbf{P}$ and $\mathbf{P}(\mathbf{y} - \mathbf{Py}) = \mathbf{0}$ by idempotence. QED

In view of Theorem 2 we call a symmetric idempotent matrix a *projection matrix*. If $\mathbf{P}$ is a projection matrix, then

$$(\mathbf{I} - \mathbf{P})^{\mathrm{T}} = \mathbf{I} - \mathbf{P}^{\mathrm{T}} = \mathbf{I} - \mathbf{P}$$

and

$$(\mathbf{I} - \mathbf{P})^2 = \mathbf{I} - 2\mathbf{P} + \mathbf{P}^2 = \mathbf{I} - \mathbf{P}$$

so $\mathbf{I} - \mathbf{P}$ is a projection matrix. Actually we can say much more.

THEOREM 3 If $\mathbf{P}$ is the projection onto L then $\mathbf{I} - \mathbf{P}$ is the projection onto $L^{\perp}$.

PROOF Let $L^{\perp} = M$: then by Theorem 4.1.2, $M^{\perp} = L$. Thus for any $\mathbf{y}$

$$\mathbf{y}_M \in L^{\perp} \quad \text{and} \quad \mathbf{y} - \mathbf{y}_M \in L$$

so $\mathbf{y}_M = \mathbf{y} - \mathbf{y}_L$. Letting $\mathbf{Q}$ be the projection onto M we have

$$\mathbf{Qy} = \mathbf{y} - \mathbf{Py} \text{ for all } \mathbf{y}$$

so $\mathbf{Q} = \mathbf{I} - \mathbf{P}$. QED

The most popular application of projections is *least-squares estimation*. Suppose we believe that imports, y, into some country can be approximately explained by national income z and an index of relative prices, r, via a simple relation of the form

$$y = \beta_1 + \beta_2 z + \beta_3 r + \varepsilon. \tag{4}$$

Here β_1, β_2, β_3 are unknown parameters and ε is an unobservable random disturbance. Suppose we have data for $n(> 3)$ time periods on y, z and r. We want to use this data to estimate β_1, β_2, β_3.

Let $\mathbf{y}$, $\mathbf{z}$, $\mathbf{r}$ denote the n-vectors of observations on y, z, r and let $\mathbf{v}$ denote the n-vector consisting entirely of 1's. Suppose that there is sufficient variation in the determining variables that $\mathbf{v}, \mathbf{z}, \mathbf{r}$ are linearly independent. Then the $n \times 3$ matrix $\mathbf{X} = (\mathbf{v}, \mathbf{z}, \mathbf{r})$ is a basis for its range, which we denote by L. Thus (4) states that $\mathbf{y}$ is equal to a linear combination of the columns of $\mathbf{X}$, plus a vector of random disturbances: the weights in the linear combination form the 3-vector $\boldsymbol{\beta} = (\beta_1, \beta_2, \beta_3)^{\mathrm{T}}$. In least-squares estimation, $\boldsymbol{\beta}$ is estimated by that 3-vector $\hat{\boldsymbol{\beta}}$ such that $\mathbf{X}\hat{\boldsymbol{\beta}}$ is the vector in L *closest* to $\mathbf{y}$. Thus $\hat{\boldsymbol{\beta}}$ is given by

$$\mathbf{X}\hat{\boldsymbol{\beta}} = \mathbf{P}\mathbf{y} = \mathbf{X}(\mathbf{X}^{\mathrm{T}}\mathbf{X})^{-1}\mathbf{X}^{\mathrm{T}}\mathbf{y}.$$

Since the columns of $\mathbf{X}$ are linearly independent, we have

$$\hat{\boldsymbol{\beta}} = (\mathbf{X}^{\mathrm{T}}\mathbf{X})^{-1}\mathbf{X}^{\mathrm{T}}\mathbf{y}. \tag{5}$$

To see why this is called least-squares estimation, let $\mathbf{b}$ be any 3-vector and consider the *residual vector*

$$\mathbf{e} = \mathbf{y} - \mathbf{X}\mathbf{b}.$$

Then $\|\mathbf{y} - \mathbf{X}\mathbf{b}\| = \sqrt{\mathbf{e}^{\mathrm{T}}\mathbf{e}}$, so choosing $\mathbf{b} = \hat{\boldsymbol{\beta}}$ is equivalent to minimising the *residual sum of squares*,

$$e_1^2 + e_2^2 + \cdots + e_n^2.$$

Letting $\hat{\boldsymbol{\varepsilon}}$ denote the least-squares residual vector $\mathbf{y} - \mathbf{X}\hat{\boldsymbol{\beta}}$, we have

$$\hat{\boldsymbol{\varepsilon}} = (\mathbf{I} - \mathbf{P})\mathbf{y}.$$

By Theorem 3, $\hat{\boldsymbol{\varepsilon}}$ is the projection of $\mathbf{y}$ onto the space $L^{\perp}$, which is the kernel of $\mathbf{X}^{\mathrm{T}}$.

To calculate the least-squares estimates there is no need to calculate $\mathbf{P}$ or even $(\mathbf{X}^{\mathrm{T}}\mathbf{X})^{-1}$: we can simply solve the *normal equations*

$$\mathbf{X}^{\mathrm{T}}\mathbf{X}\hat{\boldsymbol{\beta}} = \mathbf{X}^{\mathrm{T}}\mathbf{y}$$

by Gaussian elimination. But there are statistical purposes, such as the calculation of standard errors, for which knowledge of $(\mathbf{X}^{\mathrm{T}}\mathbf{X})^{-1}$ is useful. Since $\mathbf{X}^{\mathrm{T}}\mathbf{X}$ is positive definite we may invert it by the triangular factorisation method of the last section.

Exercises

1. Let $\mathbf{X}$ and $\mathbf{B}$ be bases for the *same* linear subspace L of R^n. Defining the matrices

$$\mathbf{P}_1 = \mathbf{X}(\mathbf{X}^{\mathrm{T}}\mathbf{X})^{-1}\mathbf{X}^{\mathrm{T}}, \quad \mathbf{P}_2 = \mathbf{B}(\mathbf{B}^{\mathrm{T}}\mathbf{B})^{-1}\mathbf{B}^{\mathrm{T}}$$

we have $\mathbf{P}_1 = \mathbf{P}_2 = \mathbf{P}$ where $\mathbf{P}$ is the projection onto L.

The fact that $\mathbf{P}_1 = \mathbf{P}_2$ can be proved from first principles without mentioning projections. Supply this proof.

2. Show that the projection onto L is the only matrix $\mathbf{P}$ such that $\mathbf{Px} = \mathbf{x}$ for all $\mathbf{x}$ in L and $\mathbf{Pw} = \mathbf{0}$ for all $\mathbf{w}$ in $L^\perp$.

3. Let L_1 and L_2 be linear subspaces of R^n such that $L_1 \subset L_2$. Let $\mathbf{P}_1$ and $\mathbf{P}_2$ denote the projections onto L_1 and L_2. Show that

$$\mathbf{P}_1\mathbf{P}_2 = \mathbf{P}_2\mathbf{P}_1 = \mathbf{P}_1$$

and that $\mathbf{P}_2 - \mathbf{P}_1$ is the projection onto $L_1^\perp \cap L_2$.

4. Let $\mathbf{X} = \begin{pmatrix} 1 & 0 & 2 \\ 1 & 1 & -2 \\ 0 & -1 & 1 \\ -1 & 0 & 0 \\ 1 & 0 & -1 \end{pmatrix}$, $\mathbf{y} = \begin{pmatrix} 3 \\ 1 \\ -2 \\ 4 \\ 5 \end{pmatrix}$.

Find the 3-vector $\mathbf{b}$ which minimises $\|\mathbf{y} - \mathbf{Xb}\|$.

4.4 Orthonormal bases

Let $\mathbf{X}$ be an $n \times k$ matrix of full column rank. Suppose we wish to calculate the projection $\mathbf{P}$ onto the range L of $\mathbf{X}$. We know that one way to do this is to compute $(\mathbf{X}^\mathrm{T}\mathbf{X})^{-1}$ and write $\mathbf{P} = \mathbf{X}(\mathbf{X}^\mathrm{T}\mathbf{X})^{-1}\mathbf{X}^\mathrm{T}$. Recall, however, that $\mathbf{P} = \mathbf{B}(\mathbf{B}^\mathrm{T}\mathbf{B})^{-1}\mathbf{B}^\mathrm{T}$ for *any* basis $\mathbf{B}$ of L. In particular, if there exists a basis $\mathbf{V}$ for L such that $\mathbf{V}^\mathrm{T}\mathbf{V} = \mathbf{I}_k$, then $\mathbf{P} = \mathbf{V}\mathbf{V}^\mathrm{T}$. We show in this section that this way of computing a projection can always be done, that the construction has convenient implications for least-squares estimation, and that it leads to the important concept of the pseudoinverse of a matrix.

Given a linear subspace L of R^n, with dimension $k > 0$, we define an *orthonormal basis* for L to be a basis $\mathbf{V}$ such that $\mathbf{V}^\mathrm{T}\mathbf{V} = \mathbf{I}_k$. To explain the name, let the columns of $\mathbf{V}$ be $\mathbf{v}_1, \ldots, \mathbf{v}_k$. Then

$$\mathbf{v}_i^\mathrm{T}\mathbf{v}_j = 0 \quad \text{for } i \neq j, \tag{1}$$

$$\|\mathbf{v}_i\| = 1 \quad \text{for all } i. \tag{2}$$

The 'ortho' comes from (1) and the 'normal' from (2). An orthonormal basis for R^n is called an *orthogonal matrix* of order n. Thus an orthogonal matrix is a nonsingular matrix whose inverse is its transpose.

The following theorem and its proof give a recipe for constructing orthonormal bases from arbitrary ones.

THEOREM 1 (Gram-Schmidt Orthogonalisation Theorem)
Let $\mathbf{X}$ be an $n \times k$ matrix of full column rank, and let its range be L. Then there exist an orthonormal basis $\mathbf{V}$ for L and a nonsingular upper triangular matrix $\mathbf{U}$ such that $\mathbf{X} = \mathbf{VU}$.

PROOF Let $\mathbf{X} = (\mathbf{x}_1, \ldots, \mathbf{x}_k)$. We construct an $n \times k$ matrix $\mathbf{Y} = (\mathbf{y}_1, \ldots, \mathbf{y}_k)$ and a $k \times k$ upper triangular matrix $\mathbf{W}$ with 1's down the diagonal, satisfying the following properties.

(i) $\mathbf{y}_i \neq \mathbf{0}$ for all i,
(ii) $\mathbf{y}_i^{\mathrm{T}}\mathbf{y}_j = 0$ if $i \neq j$,
(iii) $\mathbf{X} = \mathbf{YW}$.

Begin by setting $\mathbf{y}_1 = \mathbf{x}_1$. Then $\mathbf{y}_1 \neq \mathbf{0}$ so we can define the number

$$w_{12} = \mathbf{y}_1^{\mathrm{T}}\mathbf{x}_2/\mathbf{y}_1^{\mathrm{T}}\mathbf{y}_1.$$

Now let $\mathbf{y}_2 = \mathbf{x}_2 - w_{12}\mathbf{y}_1$. Since $w_{12}\mathbf{y}_1^{\mathrm{T}}\mathbf{y}_1 = \mathbf{y}_1^{\mathrm{T}}\mathbf{x}_2$, $\mathbf{y}_1^{\mathrm{T}}\mathbf{y}_2 = 0$. Also $\mathbf{y}_1 = \mathbf{x}_1$, so $\mathbf{y}_2 \neq \mathbf{0}$ by the linear independence of $\mathbf{x}_1$ and $\mathbf{x}_2$. We can therefore define the numbers

$$w_{13} = \mathbf{y}_1^{\mathrm{T}}\mathbf{x}_3/\mathbf{y}_1^{\mathrm{T}}\mathbf{y}_1, \quad w_{23} = \mathbf{y}_2^{\mathrm{T}}\mathbf{x}_3/\mathbf{y}_2^{\mathrm{T}}\mathbf{y}_2.$$

Let $\mathbf{y}_3 = \mathbf{x}_3 - w_{13}\mathbf{y}_1 - w_{23}\mathbf{y}_2$. Since $w_{13}\mathbf{y}_1^{\mathrm{T}}\mathbf{y}_1 = \mathbf{y}_1^{\mathrm{T}}\mathbf{x}_3$ and $\mathbf{y}_1^{\mathrm{T}}\mathbf{y}_2 = 0$, $\mathbf{y}_1^{\mathrm{T}}\mathbf{y}_3 = 0$. A similar argument shows that $\mathbf{y}_2^{\mathrm{T}}\mathbf{y}_3 = 0$. Also, $\mathbf{y}_1$ and $\mathbf{y}_2$ are linear combinations of $\mathbf{x}_1$ and $\mathbf{x}_2$, so $\mathbf{y}_3 \neq \mathbf{0}$ by the linear independence of $\mathbf{x}_1$, $\mathbf{x}_2$, $\mathbf{x}_3$. We can therefore define the numbers

$$w_{i4} = \mathbf{y}_i^{\mathrm{T}}\mathbf{x}_4/\mathbf{y}_i^{\mathrm{T}}\mathbf{y}_i \quad \text{for } i = 1, 2, 3.$$

Let $\mathbf{y}_4 = \mathbf{x}_4 - w_{14}\mathbf{y}_1 - w_{24}\mathbf{y}_2 - w_{34}\mathbf{y}_3$ and carry on. We continue until $\mathbf{y}_k$ is obtained: then $\mathbf{Y}$ and $\mathbf{W}$ satisfy (i), (ii) and (iii).

Let $\mathbf{D} = \operatorname{diag}(\|\mathbf{y}_1\|, \ldots, \|\mathbf{y}_k\|)$ and define the matrices $\mathbf{V} = \mathbf{YD}^{-1}$, $\mathbf{U} = \mathbf{DW}$. Then $\mathbf{V}^{\mathrm{T}}\mathbf{V} = \mathbf{I}_k$, $\mathbf{U}$ is upper triangular with positive diagonal entries and $\mathbf{X} = \mathbf{VU}$. This last equation and the nonsingularity of $\mathbf{U}$ imply that $\mathbf{V}$ is a basis for L. QED

Example Let $\mathbf{X}$ be the nonsingular matrix

$$\begin{pmatrix} 0 & -1 & 2 \\ 2 & 0 & -1 \\ -1 & 2 & 0 \end{pmatrix}.$$

We want to compute an orthogonal matrix $\mathbf{V}$ and an upper triangular matrix $\mathbf{U}$ such that $\mathbf{X} = \mathbf{VU}$.

We proceed as in the proof of Theorem 1. Let

$$\mathbf{y}_1 = \mathbf{x}_1 = (0 \quad 2 \quad -1)^{\mathrm{T}} \quad \text{so that } \mathbf{y}_1^{\mathrm{T}}\mathbf{y}_1 = 5.$$

Let $w_{12} = \mathbf{y}_1^{\mathrm{T}}\mathbf{x}_2/\mathbf{y}_1^{\mathrm{T}}\mathbf{y}_1 = -\frac{2}{5}$ and define

$$\begin{aligned}\mathbf{y}_2 = \mathbf{x}_2 - w_{12}\mathbf{y}_1 &= (-1 \quad 0 \quad 2)^{\mathrm{T}} + (\tfrac{2}{5})(0 \quad 2 \quad -1)^{\mathrm{T}} \\ &= (-1 \quad \tfrac{4}{5} \quad \tfrac{8}{5})^{\mathrm{T}}\end{aligned}$$

so that $\mathbf{y}_2^{\mathrm{T}}\mathbf{y}_2 = \frac{21}{5}$. Let

$$w_{13} = \mathbf{y}_1^{\mathrm{T}}\mathbf{x}_3/\mathbf{y}_1^{\mathrm{T}}\mathbf{y}_1 = -\tfrac{2}{5}, \quad w_{23} = \mathbf{y}_2^{\mathrm{T}}\mathbf{x}_3/\mathbf{y}_2^{\mathrm{T}}\mathbf{y}_2 = -\tfrac{2}{3}.$$

Then $\mathbf{y}_3 = \mathbf{x}_3 - w_{13}\mathbf{y}_1 - w_{23}\mathbf{y}_2 = (\frac{4}{3}\ \frac{1}{3}\ \frac{2}{3})^{\mathrm{T}}$, so $\mathbf{y}_3^{\mathrm{T}}\mathbf{y}_3 = \frac{7}{3}$.
We can now write down the matrices

$$\mathbf{Y} = \begin{pmatrix} 0 & -1 & \frac{4}{3} \\ 2 & \frac{4}{5} & \frac{1}{3} \\ -1 & \frac{8}{5} & \frac{2}{3} \end{pmatrix}, \quad \mathbf{W} = \begin{pmatrix} 1 & -\frac{2}{3} & -\frac{2}{5} \\ 0 & 1 & -\frac{2}{3} \\ 0 & 0 & 1 \end{pmatrix}.$$

Thus $\mathbf{V} = \mathbf{Y}\mathbf{D}^{-1}$ and $\mathbf{U} = \mathbf{D}\mathbf{W}$ where $\mathbf{D} = \operatorname{diag}(\sqrt{5}, \sqrt{\frac{21}{5}}, \sqrt{\frac{7}{3}})$.

Let us see how Theorem 1 can be applied to least-squares problems. We have an $n \times k$ matrix $\mathbf{X}$ of full column rank and an n-vector $\mathbf{y}$, and wish to compute $\hat{\boldsymbol{\beta}} = (\mathbf{X}^{\mathrm{T}}\mathbf{X})^{-1}\mathbf{X}^{\mathrm{T}}\mathbf{y}$. Let $\mathbf{V}$ and $\mathbf{U}$ be calculated as above: this is known as 'orthogonalising the data'. Clearly

$$\mathbf{V}^{\mathrm{T}}\mathbf{V} = \mathbf{I}_k, \quad \mathbf{X}^{\mathrm{T}}\mathbf{X} = \mathbf{U}^{\mathrm{T}}\mathbf{U} \quad \text{and} \quad \hat{\boldsymbol{\beta}} = \mathbf{U}^{-1}\mathbf{V}^{\mathrm{T}}\mathbf{y}.$$

Thus $\hat{\boldsymbol{\beta}}$ can be calculated by back-substitution on the triangular system

$$\mathbf{U}\hat{\boldsymbol{\beta}} = \mathbf{V}^{\mathrm{T}}\mathbf{y}. \tag{3}$$

The initial orthogonalisation of the data can be time-consuming but this way of computing least-squares estimates has numerical advantages concerning rounding errors. We shall use its logic to explain a popular procedure known as stepwise least squares (this can also be done using Gaussian elimination on the normal equations).

Let $l < k$ and let $\mathbf{X}_1$ denote the first l columns of $\mathbf{X}$. Suppose that, having computed $\hat{\boldsymbol{\beta}}$, one also wishes to compute the l-vector

$$\hat{\boldsymbol{\alpha}} = (\mathbf{X}_1^{\mathrm{T}}\mathbf{X}_1)^{-1}\mathbf{X}_1^{\mathrm{T}}\mathbf{y}.$$

(In the 'imports' example of the last section, this would correspond to experimenting with a model in which imports are affected by income but not by relative prices.) Since $\mathbf{U}$ is upper triangular we have $\mathbf{X}_1 = \mathbf{V}_1\mathbf{U}_1$ where $\mathbf{V}_1$ denotes the first l columns of $\mathbf{V}$ and $\mathbf{U}_1$ is the top-left $l \times l$ submatrix of $\mathbf{U}$. Since $\mathbf{V}^{\mathrm{T}}\mathbf{V} = \mathbf{I}_k$, $\mathbf{V}_1^{\mathrm{T}}\mathbf{V}_1 = \mathbf{I}_l$, so

$$\mathbf{U}_1\hat{\boldsymbol{\alpha}} = \mathbf{V}_1^{\mathrm{T}}\mathbf{y}. \tag{4}$$

The right-hand side of (4) is simply the first l components of the right-hand side of (3). Also $\mathbf{U}_1$ is upper triangular. So, given $\mathbf{U}$ and $\mathbf{V}$ and $\mathbf{V}^{\mathrm{T}}\mathbf{y}$, the only recomputation that needs to be done to get $\hat{\boldsymbol{\alpha}}$ is the solution of (4) by back-substitution.

In practice, $\hat{\boldsymbol{\alpha}}$ is usually computed before $\hat{\boldsymbol{\beta}}$. Since the Gram-Schmidt orthogonalisation calculates $\mathbf{U}_1$ and $\mathbf{V}_1$ on the way to computing $\mathbf{U}$ and $\mathbf{V}$, we can solve (4) for $\hat{\boldsymbol{\alpha}}$ before completing the construction of $\mathbf{U}$ and $\mathbf{V}$. Doing this for all $l = 1, \ldots, k$ is called *stepwise least squares*: statistical purists have been known to frown on this as 'data mining' but it is done all the time.

We end this chapter with a topic which has attracted much attention in recent years, notably from statisticians.

We return to more standard algebraic notation, with the typical object of interest an $m \times n$ matrix called $\mathbf{A}$ rather than an $n \times k$ matrix called $\mathbf{X}$. Let $\mathbf{A}$ be such a matrix, $\mathbf{b}$ an m-vector. Suppose that $\mathbf{A}$ has full column rank. If $m > n$ the system $\mathbf{Ax} = \mathbf{b}$ will typically have no solution: the n-vector $\mathbf{x}$ which brings $\mathbf{Ax}$ closest to $\mathbf{b}$ is the least-squares estimate $(\mathbf{A}^{\mathrm{T}}\mathbf{A})^{-1}\mathbf{A}^{\mathrm{T}}\mathbf{b}$. Thus the matrix

$$\mathbf{A}^{+} = (\mathbf{A}^{\mathrm{T}}\mathbf{A})^{-1}\mathbf{A}^{\mathrm{T}}$$

is the nearest thing to an inverse that $\mathbf{A}$ possesses. Notice that if $\mathbf{b}$ happens to be in rng $\mathbf{A}$, then $\mathbf{A}^{+}\mathbf{b}$ is indeed the unique solution to the system; and if $m = n$ then $\mathbf{A}^{+} = \mathbf{A}^{-1}$.

It turns out that one can generalise $\mathbf{A}^{+}$ to cases where $\mathbf{A}$ does not have full column rank, including the case where $\mathbf{A}$ is square and singular.

THEOREM 2 Let $\mathbf{A}$ be any $m \times n$ matrix and let $\mathbf{P}$ and $\mathbf{Q}$ be the projections onto rng $\mathbf{A}$ and rng $\mathbf{A}^{\mathrm{T}}$ respectively (so $\mathbf{P}$ is $m \times m$ and $\mathbf{Q}$ is $n \times n$). Then there exists exactly one $n \times m$ matrix $\mathbf{A}^{+}$ with the following properties:
(i) rk $\mathbf{A}^{+}$ = rk $\mathbf{A}$;
(ii) $\mathbf{AA}^{+} = \mathbf{P}$;
(iii) $\mathbf{A}^{+}\mathbf{A} = \mathbf{Q}$.

PROOF Let rk $\mathbf{A} = r$. If $r = 0$ then $\mathbf{A} = \mathbf{O}_{mn}$ and $\mathbf{A}^{+} = \mathbf{O}_{nm}$. So suppose from now on that $r > 0$. Let $\mathbf{V}$ and $\mathbf{W}$ be orthonormal bases for rng $\mathbf{A}$ and rng $\mathbf{A}^{\mathrm{T}}$ respectively: such exist by Theorem 1. Then

$$\mathbf{VV}^{\mathrm{T}} = \mathbf{P}, \quad \mathbf{WW}^{\mathrm{T}} = \mathbf{Q} \quad \text{and} \quad \mathbf{V}^{\mathrm{T}}\mathbf{V} = \mathbf{W}^{\mathrm{T}}\mathbf{W} = \mathbf{I}_r.$$

Since $\mathbf{Py} = \mathbf{y}$ for every $\mathbf{y}$ in rng $\mathbf{A}$, $\mathbf{PA} = \mathbf{A}$. Similarly $\mathbf{QA}^{\mathrm{T}} = \mathbf{A}^{\mathrm{T}}$, so $\mathbf{AQ} = \mathbf{A}$ by symmetry of $\mathbf{Q}$. Thus

$$\mathbf{A} = \mathbf{PAQ} = \mathbf{VCW}^{\mathrm{T}} \quad \text{where } \mathbf{C} = \mathbf{V}^{\mathrm{T}}\mathbf{AW}.$$

Now $\mathbf{V}$ has r columns and full column rank, and the same holds for $\mathbf{W}$. So by Theorem 2.4.5, the $r \times r$ matrix $\mathbf{C}$ is nonsingular. Letting $\mathbf{A}^{+} = \mathbf{WC}^{-1}\mathbf{V}^{\mathrm{T}}$ and applying Theorem 2.4.5 again, we see that $\mathbf{A}^{+}$ has rank r. Also

$$\mathbf{AA}^{+} = \mathbf{VC}(\mathbf{W}^{\mathrm{T}}\mathbf{W})\mathbf{C}^{-1}\mathbf{V}^{\mathrm{T}} = \mathbf{V}(\mathbf{CC}^{-1})\mathbf{V}^{\mathrm{T}} = \mathbf{VV}^{\mathrm{T}} = \mathbf{P}$$

and $\mathbf{A}^{+}\mathbf{A} = \mathbf{Q}$ by a similar argument.

To prove uniqueness, let $\mathbf{A}^{+}$ be any matrix satisfying (i), (ii) and (iii); we wish to show that $\mathbf{A}^{+} = \mathbf{WC}^{-1}\mathbf{V}^{\mathrm{T}}$. Let $\mathbf{B} = (\mathbf{A}^{+})^{\mathrm{T}}$. By (i) and the Rank Theorem, rk $\mathbf{B}$ = rk $\mathbf{A}$. By transposition of (ii), the range of $\mathbf{B}$ contains the range of $\mathbf{P}$, which is rng $\mathbf{A}$. Hence by Theorem 2.3.2, rng $\mathbf{B}$ = rng $\mathbf{A}$. A similar argument using (i) and (iii) shows that rng $\mathbf{A}^{+}$ = rng $\mathbf{A}^{\mathrm{T}}$. Hence

$$\mathbf{A}^{+} = \mathbf{QA}^{+}\mathbf{P} = \mathbf{WGV}^{\mathrm{T}} \quad \text{where } \mathbf{G} = \mathbf{W}^{\mathrm{T}}\mathbf{A}^{+}\mathbf{V}.$$

Now (ii) may be written

$$(\mathbf{VCW}^{\mathrm{T}})(\mathbf{WGV}^{\mathrm{T}}) = \mathbf{VV}^{\mathrm{T}}$$

whence $\mathbf{V}(\mathbf{CG} - \mathbf{I}_r)\mathbf{V}^{\mathrm{T}} = \mathbf{0}$. Since $\mathbf{V}$ has full column rank, $\mathbf{CG} = \mathbf{I}$. Thus $\mathbf{A}^{+} = \mathbf{WC}^{-1}\mathbf{V}^{\mathrm{T}}$. QED

The matrix $\mathbf{A}^{+}$ of Theorem 2 is called the *pseudoinverse* (or *Moore–Penrose generalised inverse*) of $\mathbf{A}$. Notice that the uniqueness part of the proof shows that

$$\text{rng } (\mathbf{A}^{+})^{\mathrm{T}} = \text{rng } \mathbf{A} \quad \text{and} \quad \text{rng } \mathbf{A}^{+} = \text{rng } \mathbf{A}^{\mathrm{T}}.$$

It follows that

$$(\mathbf{A}^{+})^{+} = \mathbf{A} \quad \text{and} \quad (\mathbf{A}^{\mathrm{T}})^{+} = (\mathbf{A}^{+})^{\mathrm{T}}.$$

This is a sensible definition of a generalised inverse, for two reasons. It reduces to the right sort of thing in special cases, and it has inverse-like properties in the general case.

To take the first point first, suppose $\mathbf{A}$ has full column rank. Then $\mathbf{P} = \mathbf{A}(\mathbf{A}^{\mathrm{T}}\mathbf{A})^{-1}\mathbf{A}^{\mathrm{T}}$ and the only matrix $\mathbf{A}^{+}$ for which $\mathbf{AA}^{+} = \mathbf{P}$ is given by $\mathbf{A}^{+} = (\mathbf{A}^{\mathrm{T}}\mathbf{A})^{-1}\mathbf{A}^{\mathrm{T}}$. In particular, $\mathbf{A}^{+} = \mathbf{A}^{-1}$ if $\mathbf{A}$ is square and nonsingular.

Now look at the general case. First, we have the reflexive property

$(\mathbf{A}^+)^+ = \mathbf{A}$. Second, $\mathbf{AA}^+ = \mathbf{P}$ means that $\mathbf{AA}^+$ operates like an identity matrix on the range of $\mathbf{A}$. Third, $\mathbf{A}^+\mathbf{A} = \mathbf{Q}$ means that $\mathbf{A}^+\mathbf{A}$ acts like an identity matrix on the range of $\mathbf{A}^\mathrm{T}$.

A final note about the construction of pseudoinverses. In the proof of Theorem 2 we chose arbitrary orthonormal bases $\mathbf{V}$ and $\mathbf{W}$ for rng $\mathbf{A}$ and rng $\mathbf{A}^\mathrm{T}$ respectively, expressed $\mathbf{A}$ in the form $\mathbf{VCW}^\mathrm{T}$ and let $\mathbf{A}^+ = \mathbf{WC}^{-1}\mathbf{V}^\mathrm{T}$. It may be shown that $\mathbf{V}$ and $\mathbf{W}$ can always be chosen in such a way that $\mathbf{C}$ is a diagonal matrix. This is known as the *singular value decomposition* of $\mathbf{A}$: its construction requires eigenvalue theory, which is not treated in this book.

Exercises

1. The *trace* of a square matrix is the sum of its diagonal entries. Show that if $\mathbf{A}$ is $m \times n$ and $\mathbf{B}$ is $n \times m$, the square matrices $\mathbf{AB}$ and $\mathbf{BA}$ have the same trace. Hence show that the trace of a projection matrix is its rank.

2. Repeat Exercise 4.3.4 using orthogonalisation.

3. Let $\mathbf{A}$ be any matrix. Show that $\mathbf{A}^+$ is the only matrix $\mathbf{B}$ such that $\mathbf{AB}$ and $\mathbf{BA}$ are symmetric matrices and $\mathbf{ABA} = \mathbf{A}$ and $\mathbf{BAB} = \mathbf{B}$.

4. Consider the least-squares problem

   $$\text{minimise } \|\mathbf{y} - \mathbf{Xb}\|, \quad \text{given } \mathbf{X} \text{ and } \mathbf{y}$$

   where $\mathbf{X}$ does not have full column rank. Show that there are infinitely many solutions for $\mathbf{b}$: these are given by

   $$\mathbf{b} = \mathbf{X}^+\mathbf{y} + \mathbf{w}$$

   where $\mathbf{w}$ is an arbitrary member of the kernel of $\mathbf{X}$.

5 Introduction to Linear Programming

Mathematical Programming is that branch of applied mathematics which concerns itself with the maximisation or minimisation of functions of several variables, subject to constraints on those variables.

Some readers will be familiar already with one method of maximisation subject to constraints, namely the Lagrange-multiplier method. An obvious difficulty with this method is that it deals only with constraints in the form of equations, whereas real-world maximisation problems and problems of economic theory typically involve constraints in the form of inequalities. For example, a corporate planner will know that his firm has a credit line at a bank but there may be no *a priori* reason why the firm should borrow to the limit of its facility. At a more theoretical level, the *homo oeconomicus* who maximises 'utility' of bundles of bread and cheese will typically be faced with non-negativity constraints as well as a budget constraint: he cannot consume negative amounts of bread or of cheese! This gives rise to the 'corner optima' discussed in price-theory textbooks.

The Lagrange-multiplier theory can be extended to the case of inequality constraints. The main theorem in this area, the Kuhn–Tucker Theorem, will be discussed in Chapter 9. There are, however, two problems associated with the application of the Lagrange-multiplier technique and its extension by Kuhn and Tucker.

First, these results do not yield immediately methods for actually *calculating* an optimum, in any but the simplest cases.

Second, calculus methods are not good at dealing with the existential issue of whether a given problem has a solution at all. Such questions cannot be regarded as academic (in the pejorative sense). Consider, for example, a development planner. He may have talked to representatives of so many interest groups that he has built into his model a great mass of constraints which simply cannot be satisfied simultaneously. At the other extreme, he may have modelled his problem in so inept a way that the mathematical maximum is effectively infinity: in other words, he has under-constrained his model. It may well not be obvious from the data that both these traps have been avoided.

In Chapters 5–8, we shall be concerned with the special case of mathematical programming in which all relevant functions are assumed to be linear. The beauty of linear programming is that it is strong precisely where calculus methods are weak: it does have simple computational algorithms and a fully-developed existence theory which is integrated with the algorithms.

This first chapter on linear programming is concerned entirely with groundwork: we define terms and classify cases. Only at the end of Section 5.3 do we indicate our line of attack for solving linear programmes.

5.1 Linear programmes

A *linear form* in n variables $x_1, \ldots, x_n$ is an expression of the form

$$a_1 x_1 + \cdots + a_n x_n$$

where $a_1, \ldots, a_n$ are given numbers. A *linear constraint* on the variables $x_1, \ldots, x_n$ is a statement

either of the form $f(x_1, \ldots, x_n) \leq b$
or of the form $f(x_1, \ldots, x_n) \geq b$
or of the form $f(x_1, \ldots, x_n) = b$

where in each case f is a given linear form and b a given number.

A *linear maximisation programme* is a problem in which one attempts to maximise a given linear form in variables $x_1, \ldots, x_n$, subject to a finite number of linear constraints on those variables. A *linear minimisation programme* is similar, save that 'maximise' is replaced by 'minimise'. In each case, the linear form to be maximised or minimised is called the *objective function* of the programme.

We shall illustrate these concepts with two examples. Example 1 is a simple and stylised 'diet problem': the use of linear programming in solving such problems has been common in agricultural economics since the 1940s. Example 2, again very simple and stylised, illustrates how problems in production planning may be fitted into the linear programming framework. For both examples, we are concerned in this section with problem formulation only: solutions will be computed in Section 6.2.

Example 1 Suppose that one unit of bread contains 3 units of carbohydrates, 4 units of vitamins and 1 unit of protein and costs 25 units of money. Suppose also that one unit of cheese contains 1 unit of carbohydrate, 3 units of vitamins and 2 units of protein, and costs 40 units of money. You wish to buy the least costly bundle of bread and cheese compatible with the nutritional requirements of *at least* 8 units of carbohydrates, *at least* 19 units of vitamins and *at least* 7 units of protein. How much bread and how much cheese should you buy?

To formulate this as a linear programme, let p_1 and p_2 denote respectively the amounts of bread and cheese bought: here, as throughout our discussion of linear programming, we do *not* restrict our variables to be integers. The quantities of carbohydrates, vitamins and protein provided by this bundle are respectively

$$3p_1 + p_2, \quad 4p_1 + 3p_2 \quad \text{and} \quad p_1 + 2p_2.$$

The nutritional constraints are therefore

$$\begin{aligned} 3p_1 + p_2 &\geq 8 && \text{for carbohydrates} \\ 4p_1 + 3p_2 &\geq 19 && \text{for vitamins} \\ p_1 + 2p_2 &\geq 7 && \text{for protein.} \end{aligned}$$

The remaining constraints are that p_1 and p_2 be non-negative. The cost of the bundle is $25p_1 + 40p_2$ units of money.

Your problem is therefore to

$$\begin{aligned} \text{minimise} \quad & 25p_1 + 40p_2 \\ \text{subject to} \quad & 3p_1 + p_2 \geq 8 \\ & 4p_1 + 3p_2 \geq 19 \\ & p_1 + 2p_2 \geq 7 \\ & p_1 \geq 0, p_2 \geq 0. \end{aligned}$$

This is a linear minimisation programme in 2 variables with 5 constraints.

Example 2 A firm has four machines, M_1, M_2, M_3, M_4, which can be used for up to 60, 60, 50, 65 hours per week, respectively. The only direct cost is labour at £4 per hour. The firm can produce 'widgets' by either of two techniques, A and B. No cost is involved in switching from one technique to the other in the course of a week. With technique A, production of each widget requires 3 hours on M_1, 2 on M_2, 1 on M_3 and 70 man-hours. With technique B, production of each widget requires 1 hour on M_2, 2 on M_3, 3 on M_4 and 75 man-hours. Also, 4 hours on M_4 and 20 man-hours convert a widget into a superwidget. Widgets sell at £400 each, superwidgets at £488 each. How should the firm organise production so as to maximise profits?

Here we have some choice as to what quantities to use as variables in our formulation. One reasonable procedure is to work with three variables x_1, x_2, x_3, where $x_1 + x_2$ is the total number of widgets produced per week (including those subsequently turned into super-widgets) and x_3 is the number of superwidgets produced per week; of the $x_1 + x_2$ widgets, x_1 are produced via A and x_2 via B. Clearly x_1, x_2 and x_3 must be non-negative. Also each superwidget must start life as a widget, so we have the throughput constraint

$$x_3 \leq x_1 + x_2 .$$

The remaining constraints concern machine time. The total numbers of operating hours per week on each machine are as follows:

M_1	M_2	M_3	M_4
$3x_1$	$2x_1 + x_2$	$x_1 + 2x_2$	$3x_2 + 4x_3$

Thus our time constraints are

$$\begin{aligned}
3x_1 \qquad\qquad\qquad &\leq 60 \quad (M_1)\\
2x_1 + x_2 \qquad\qquad &\leq 60 \quad (M_2)\\
x_1 + 2x_2 \qquad\qquad &\leq 50 \quad (M_3)\\
3x_2 + 4x_3 &\leq 65 \quad (M_4).
\end{aligned}$$

Now consider the objective function. In £ per week,

$$\begin{aligned}
\text{revenue} &= 400(x_1 + x_2 - x_3) + 488x_3 &&= 400x_1 + 400x_2 + 88x_3\\
\text{and cost} &= 4(70x_1 + 75x_2 + 20x_3) &&= 280x_1 + 300x_2 + 80x_3\\
\text{so profit} &= \text{revenue} - \text{cost} &&= 120x_1 + 100x_2 + 8x_3 .
\end{aligned}$$

Before putting all this together, let us check if any of the constraints are obviously redundant, in the sense of being entailed by the other constraints. This is indeed so. For if we add the time constraints for M_1 and M_3 and divide by 2, we see that $2x_1 + x_2 \leq 55$. Since

$55 < 60$, the time constraint for M_2 is satisfied whenever those for M_1 and M_3 are, and therefore need not be written down as a separate constraint: notice that this implies that M_2 must be idle for at least 5 hours per week. Thus the constraints we need to worry about are the time constraints for M_1, M_3 and M_4, the throughput constraint and the non-negativity constraints.

The firm's problem can now be written as follows:

$$\begin{array}{ll} \text{maximise} & 120x_1 + 100x_2 + 8x_3 \\ \text{subject to} & x_1 \leq 20 \\ & x_1 + 2x_2 \leq 50 \\ & 3x_2 + 4x_3 \leq 65 \\ & -x_1 - x_2 + x_3 \leq 0 \\ & x_1 \geq 0,\ x_2 \geq 0,\ x_3 \geq 0. \end{array}$$

This is a linear minimisation programme in 3 variables with 7 constraints.

A linear programme in *standard form* is a problem of the following type:

$$\begin{array}{ll} \text{maximise} & c_1x_1 + c_2x_2 + \cdots + c_nx_n \\ \text{subject to} & a_{11}x_1 + a_{12}x_2 + \cdots + a_{1n}x_n \leq b_1 \\ & a_{21}x_1 + a_{22}x_2 + \cdots + a_{2n}x_n \leq b_2 \\ & \vdots \\ & a_{m1}x_1 + a_{m2}x_2 + \cdots + a_{mn}x_n \leq b_m \\ & x_1 \geq 0,\ x_2 \geq 0,\ \ldots,\ x_n \geq 0. \end{array}$$

Thus Example 2 is a standard form linear programme, while Example 1 is not.

We now show that *any linear programme can be expressed in standard form.* This important result follows from the following four remarks, (i)–(iv).

(i) Any linear minimisation programme can be expressed as a linear maximisation programme. For suppose we wish to minimise the linear form $f(x_1, \ldots, x_n)$ subject to given constraints on the variables $x_1, \ldots, x_n$. This is equivalent to maximising $-f(x_1, \ldots, x_n)$ subject to the same constraints.

(ii) The constraint $f(x_1, \ldots, x_n) \geq b$ can be written

$$-f(x_1, \ldots, x_n) \leq -b.$$

(iii) Any equation constraint can be written as two weak inequality constraints: $f(x_1, \ldots, x_n) = b$ if and only if $f(x_1, \ldots, x_n) \leq b$ and $-f(x_1, \ldots, x_n) \leq -b$.

(iv) Any linear programme may be expressed as a linear programme in non-negative variables. Suppose we have a linear programme in $(r + s + t)$ variables

$y_1, \ldots, y_r$ (required to be non-negative),
$y_{r+1}, \ldots, y_{r+s}$ (required to be non-positive),
and $y_{r+s+1}, \ldots, y_{r+s+t}$ (unrestricted in sign).

We can formulate the programme in terms of $(r + s + 2t)$ non-negative variables

$$x_1, \ldots, x_r, \quad x_{r+1}, \ldots, x_{r+s},$$
$$x_{r+s+1}, \ldots, x_{r+s+t}, \quad x_{r+s+t+1}, \ldots, x_{r+s+2t}$$

by setting

$$\begin{aligned} y_j &= x_j && \text{for } j = 1, \ldots, r, \\ y_j &= -x_j && \text{for } j = r+1, \ldots, r+s, \\ y_j &= x_j - x_{j+t} && \text{for } j = r+s+1, \ldots, r+s+t. \end{aligned}$$

The non-negativity of x_j for $j = 1, \ldots, r + s$ follows from the sign restrictions on the corresponding y_j. The restriction

$$x_j \geq 0 \quad \text{for } j = r + s + 1, \ldots, r + s + 2t$$

may be made *without loss of generality* since any real number can be expressed as the difference between two non-negative numbers (for example, $2 = 2 - 0$ and $-3 = 1 - 4$).

From now on, variables which are unrestricted in sign will be called *free* variables.

Example 3 Express in standard form the linear programme

$$\begin{aligned} \text{minimise} \quad & 4y_1 + 3y_2 - 2y_3 + 4y_4 \\ \text{subject to} \quad & y_1 + 8y_2 + 5y_3 - y_4 \leq 1 \\ & y_1 - 2y_2 + 7y_3 + y_4 = 1 \\ & -y_1 + 5y_2 - y_3 - y_4 \geq 1 \\ & y_1 \geq 0,\ y_2 \text{ free},\ y_3 \geq 0,\ y_4 \leq 0. \end{aligned}$$

Setting

$$y_1 = x_1,\ y_2 = x_2 - x_3,\ y_3 = x_4,\ y_4 = -x_5$$

we have the standard form

$$\begin{aligned} \text{maximise} \quad & -4x_1 - 3x_2 + 3x_3 + 2x_4 + 4x_5 \\ \text{subject to} \quad & x_1 + 8x_2 - 8x_3 + 5x_4 + x_5 \leq 1 \\ & x_1 - 2x_2 + 2x_3 + 7x_4 - x_5 \leq 1 \\ & -x_1 + 2x_2 - 2x_3 - 7x_4 + x_5 \leq -1 \\ & x_1 - 5x_2 + 5x_3 + x_4 - x_5 \leq -1 \\ & x_1 \geq 0,\ x_2 \geq 0,\ x_3 \geq 0,\ x_4 \geq 0,\ x_5 \geq 0. \end{aligned}$$

Exercises

1. Formulate the following as a linear programme.
   A firm has 3 workshops, each available for up to 50 hours per week. It produces two products, A and B. A sells at £65 per unit, B at £54 per unit. Production of each unit of A requires 5 hours in workshop I, 3 hours in workshop II, 7 hours in workshop III and 10 man-hours. Production of each unit of B requires 2 hours in I, 4 hours in II, 1 hour in III and 8 man-hours. The only direct cost is labour at £3 per hour. The firm wishes to maximise profits.

2. Express the diet problem of Example 1 in standard form.

5.2 Feasibility and solubility

Suppose we have a linear maximisation programme, not necessarily in standard form, with objective function $f(x_1, \ldots, x_n)$. Such a programme is said to be *feasible* if it is possible to satisfy the constraints, *infeasible* otherwise. Given that the programme is feasible, there may or may not exist a number a such that

$$f(x_1, \ldots, x_n) \leq a \quad \text{whenever } x_1, \ldots, x_n \text{ satisfy the constraints.}$$

If such an a exists, our programme is said to be *feasible and bounded*; if not, the programme is said to be *feasible and unbounded*.

Thus any linear maximisation programme falls into one and only one of three categories: either it is infeasible, or it is feasible and unbounded or it is feasible and bounded.

Example 1 Maximise x_1 subject to $x_1 + x_2 = 1$ and $x_1 + x_2 = 2$.
Clearly we cannot choose numbers x_1 and x_2 such that $x_1 + x_2$ is simultaneously 1 and 2. Thus this programme is *infeasible*.

Example 2 Maximise $3x_1 + 4x_2$ subject to

$$-x_1 + 2x_2 \leq 1, \quad 2x_1 - x_2 \leq -2, \quad x_1 \geq 0, \quad x_2 \geq 0.$$

Adding the first two constraints we have $x_1 + x_2 \leq -1$ which is incompatible with the non-negativity constraints. Thus this programme is *infeasible*.

Example 3 Maximise x_1 subject to

$$x_1 - 2x_2 \leq 1, \quad x_1 \geq 0, \quad x_2 \geq 0.$$

Let M be any positive number. Setting $x_1 = x_2 = M$, we see that all constraints are satisfied and the objective function takes the value M. This can be done for any positive M, however large, so we may make the objective function take a value as large as we like without violating the constraints. The programme is therefore *feasible and unbounded*.

Example 4 Maximise $2x_1 + x_2$ subject to

$$3x_1 + x_2 \leq 1 \quad \text{and} \quad x_1 + x_2 \geq 0.$$

Again let M be any positive number. By setting

$$x_1 = -M, \quad x_2 = 3M$$

we can make the objective function take the value M while satisfying the constraints. Since this is so for arbitrarily large M, the programme is *feasible and unbounded*.

Example 5 Maximise $x_1 + x_2$ subject to

$$3x_1 + 5x_2 = 3 \quad \text{and} \quad x_1 - x_2 \leq 5.$$

Set $\lambda = x_2/3$. Then $x_2 = 3\lambda$, and $x_1 = 1 - 5\lambda$ by the first constraint. We therefore want to

$$\text{maximise } 1 - 2\lambda \quad \text{subject to } 1 - 8\lambda \leq 5,$$

which is the same problem as to

$$\text{minimise } \lambda \text{ subject to } 2\lambda \geq -1.$$

Obviously we cannot do better than to take $\lambda = -\frac{1}{2}$, in which case

$$x_1 = \tfrac{7}{2}, \quad x_2 = -\tfrac{3}{2} \quad \text{and} \quad x_1 + x_2 = 2.$$

The programme is *feasible and bounded*.

We now discuss the 'solubility' of linear maximisation programmes. Again suppose that we have a linear maximisation programme with objective function $f(x_1, \ldots, x_n)$. This programme is said to be *soluble* if there exists a number π such that

$$f(x_1, \ldots, x_n) \leq \pi \text{ for } \textit{all } x_1, \ldots, x_n \text{ satisfying the constraints}$$

and $f(x_1^0, \ldots, x_n^0) = \pi$ for *some* $x_1^0, \ldots, x_n^0$ satisfying the constraints.

In this case π is called *the solution value* to the programme and

$$x_1 = x_1^0, \ldots, x_n = x_n^0$$

is said to be *a solution* to the programme.

The linear maximisation programme of Example 5 is soluble with solution value 2. A solution (in fact the only one) is

$$x_1 = \tfrac{7}{2}, \quad x_2 = -\tfrac{3}{2}.$$

The reason for the terms '*a* solution' and '*the* solution value' is that a soluble linear maximisation programme may have more than one solution: this is illustrated in Example 6 below. The solution *value* is, however, unique, being *the* greatest value which the objective function can attain, given the constraints. All we are really saying here is that there may be more than one tallest man in Cardiff but, if one of these men is exactly seven feet tall, so are they all.

Example 6

$$\begin{array}{ll} \text{Maximise} & x_1 + x_2 + x_3 - x_4 \\ \text{subject to} & 3x_1 + 2x_2 + 2x_3 - 2x_4 \leq 6 \\ & -x_1 + 3x_2 + x_3 - 5x_4 \leq 1 \\ & x_1 \geq 0, \quad x_2 \geq 0, \quad x_3 \geq 0, \quad x_4 \geq 0. \end{array}$$

Denote the objective function by z. The first constraint may then be written $x_1 + 2z \leq 6$. Since $x_1 \geq 0$, we must have $z \leq 3$. But when $x_1 = 0$, $x_2 = 3$ and $x_3 = x_4 \geq 2$, all constraints are satisfied and $z = 3$.

Thus this programme is soluble with solution value 3. A solution is

$$x_1 = 0, \quad x_2 = 3, \quad x_3 = 2, \quad x_4 = 2.$$

Another solution is

$$x_1 = 0, \quad x_2 = 3, \quad x_3 = 100, \quad x_4 = 100.$$

Up to this point, this section has been concerned with linear *maximisation* programmes. But it is easy to see how feasibility, boundedness and solubility can be defined for linear minimisation programmes. Suppose we wish to *minimise* a linear form $g(x_1, \ldots, x_n)$ subject to given constraints. Denote this programme by P, and let $\tilde{P}$ denote the equivalent maximisation programme

$$\text{maximise } -g(x_1, \ldots, x_n) \text{ subject to the constraints of } P.$$

Then P is said to be infeasible, feasible-and-unbounded or feasible-and-bounded, according as $\tilde{P}$ is infeasible, feasible-and-unbounded or feasible-and-bounded. In particular, P is feasible and unbounded if and only if it is possible to make the objective function g take a sequence of values approaching $-\infty$ without violating the constraints.

Similarly we say that P is soluble if $\tilde{P}$ is soluble. We say that

$$x_1 = x_1^0, \ldots, x_n = x_n^0$$

is a solution to P if the same values of the variables give a solution to $\tilde{P}$; and we say that π is the solution value to P if $-\pi$ is the solution value to $\tilde{P}$. Thus P is soluble with solution value π if and only if

$$g(x_1, \ldots, x_n) \geq \pi \text{ for all } x_1, \ldots, x_n \text{ satisfying the constraints}$$

and $g(x_1^0, \ldots, x_n^0) = \pi$ for some $x_1^0, \ldots, x_n^0$ satisfying the constraints.

So much for definitions. We must now discuss the relationship between feasibility and boundedness, and solubility. It is obvious that any soluble linear programme is feasible and bounded. Also the two examples given above of feasible and bounded linear programmes, namely Examples 5 and 6, were soluble. But is it true that *any* linear programme which is feasible and bounded is soluble?

The answer is 'Yes': this result, which we call the *Existence Theorem of Linear Programming*, will be proved in Section 7.1. The main point to notice here is that the Existence Theorem does require proof: it is not an immediate consequence of the definitions of this section. This point may be emphasised by the following three remarks.

(i) If we were to allow *strict* inequality constraints, the Existence Theorem would no longer hold. For consider the problem

$$\text{maximise } x \text{ subject to } x < 1.$$

This problem is feasible and bounded but it is not soluble. For given $x = 0.9$ we can do better by taking $x = 0.99$, better still by taking $x = 0.999$ and so on. But taking $x = 1$ violates the constraint.

(ii) If we were to allow nonlinear objective functions, the Existence Theorem would no longer hold. For consider the problem

$$\text{maximise } x/(1 + x) \text{ subject to } x \geq 0.$$

This is obviously feasible; it is also bounded, since $x/(1 + x) \leq 1$ for all non-negative x. On the other hand, this problem is not soluble: for

we can make the objective function take values as close as we like to 1 by making x sufficiently large but there is no number x such that $x/(1+x)=1$.

(iii) Some readers may be familiar with the Weierstrass Maximum Value Theorem, which states that a continuous function defined on a closed and bounded set in R^n is bounded and attains its bounds. (The terms 'bounded set' and 'closed set' will be defined in Section 9.1.) The Existence Theorem of linear programming is *not* an immediate consequence of the Weierstrass theorem, for the simple reason that solubility of a linear programme does not require that the set defined by the constraints be bounded. Example 6 illustrates this, as does the trivial programme

$$\text{maximise } x \text{ subject to } x \leq 0.$$

5.3 The optimality condition

From now on we shall formulate linear programmes in matrix algebra. To do this we need a vector notation for weak inequalities.

Given n-vectors $\mathbf{x}$ and $\mathbf{y}$ we write $\mathbf{x} \geq \mathbf{y}$ if $x_j \geq y_j$ for all $j = 1, \ldots, n$. In particular, $\mathbf{x} \geq \mathbf{0}$ means that every component of $\mathbf{x}$ is non-negative: such an $\mathbf{x}$ is called a *non-negative* n-vector. The statement $\mathbf{y} \leq \mathbf{x}$ means $\mathbf{x} \geq \mathbf{y}$.

The statements $\mathbf{x} \geq \mathbf{y}$ and $\mathbf{x} \leq \mathbf{y}$ are true simultaneously if and only if $\mathbf{x} = \mathbf{y}$. For some $\mathbf{x}$ and $\mathbf{y}$, both statements are false. For example, if

$$\mathbf{x} = \begin{pmatrix} 2 \\ -1 \\ 4 \end{pmatrix}, \quad \mathbf{y} = \begin{pmatrix} 3 \\ -2 \\ 1 \end{pmatrix}, \quad \mathbf{z} = \begin{pmatrix} 1 \\ -3 \\ 1 \end{pmatrix}$$

then $\mathbf{x} \geq \mathbf{z}$ and $\mathbf{z} \leq \mathbf{y}$ but we have neither $\mathbf{x} \geq \mathbf{y}$ nor $\mathbf{x} \leq \mathbf{y}$.

We state for the record the main properties of vector inequalities. Proofs are trivial.

If $\mathbf{x} \geq \mathbf{y}$ and $\mathbf{y} \geq \mathbf{z}$ then $\mathbf{x} \geq \mathbf{z}$.
If $\mathbf{x} \geq \mathbf{y}$ and $\mathbf{a} \geq \mathbf{b}$ then $\mathbf{x} + \mathbf{a} \geq \mathbf{y} + \mathbf{b}$.
If $\mathbf{x} \geq \mathbf{y}$ and $\lambda \geq 0$ then $\lambda\mathbf{x} \geq \lambda\mathbf{y}$.
If $\mathbf{x} \geq \mathbf{y}$ then $-\mathbf{y} \geq -\mathbf{x}$.
If $\mathbf{x} \geq \mathbf{y}$ and $\mathbf{z} \geq \mathbf{0}$ then $\mathbf{x}^T\mathbf{z} \geq \mathbf{y}^T\mathbf{z}$.

The generic standard-form linear programme introduced in Section

5.1 can now be written in the form

(S) maximise $\mathbf{c}^T\mathbf{x}$ subject to $\mathbf{Ax} \le \mathbf{b}$ and $\mathbf{x} \ge \mathbf{0}$

where $\mathbf{A}$ is an $m \times n$ matrix, $\mathbf{b}$ an m-vector and $\mathbf{c}$ an n-vector.

With the programme (S) we associate another linear programme called the *dual* of (S). This is the following minimisation problem in m variables $p_1, \ldots, p_m$:

(D) minimise $\mathbf{b}^T\mathbf{p}$ subject to $\mathbf{A}^T\mathbf{p} \ge \mathbf{c}$ and $\mathbf{p} \ge \mathbf{0}$.

We introduce the dual at this stage because (as we shall explain at the end of this section) the usual method of solving (S) solves (D) at the same time. This is useful for two reasons. First, some linear programmes such as diet problems occur naturally in the form (D) rather than (S). Second, it will be shown in Section 7.2 that the duals of production planning problems have an interesting economic interpretation in terms of marginal profitability.

Example In Example 5.2.6 we considered the standard-form programme

$$\begin{aligned}
\text{maximise}\quad & x_1 + x_2 + x_3 - x_4 \\
\text{subject to}\quad & 3x_1 + 2x_2 + 2x_3 - 2x_4 \le 6 \\
& -x_1 + 3x_2 + x_3 - 5x_4 \le 1 \\
& x_1 \ge 0, \quad x_2 \ge 0, \quad x_3 \ge 0, \quad x_4 \ge 0.
\end{aligned}$$

The dual of this programme is:

$$\begin{aligned}
\text{minimise}\quad & 6p_1 + p_2 \\
\text{subject to}\quad & 3p_1 - p_2 \ge 1 \\
& 2p_1 + 3p_2 \ge 1 \\
& 2p_1 + p_2 \ge 1 \\
& -2p_1 - 5p_2 \ge -1 \\
& p_1 \ge 0, \quad p_2 \ge 0.
\end{aligned}$$

We say that an n-vector $\mathbf{x}$ is *feasible* for (S) if it satisfies the constraints $\mathbf{Ax} \le \mathbf{b}$ and $\mathbf{x} \ge \mathbf{0}$. Let X denote the set of such vectors: X is non-empty if and only if (S) is feasible. The set X is what is known as a *convex set* in R^n: if $\mathbf{x}_1$ and $\mathbf{x}_2$ are members of X and $0 < \alpha < 1$ then

$$\alpha\mathbf{x}_1 + (1 - \alpha)\mathbf{x}_2 \in X.$$

To prove this, let $\mathbf{x}_1$, $\mathbf{x}_2$ and α have the properties just stated and let

$\mathbf{x} = \alpha\mathbf{x}_1 + (1-\alpha)\mathbf{x}_2$. Since α and $1-\alpha$ are positive, $\mathbf{x} \geq \mathbf{0}$. Also

$$\mathbf{b} - \mathbf{A}\mathbf{x} = \alpha(\mathbf{b} - \mathbf{A}\mathbf{x}_1) + (1-\alpha)(\mathbf{b} - \mathbf{A}\mathbf{x}_2) \geq \mathbf{0}.$$

Thus $\mathbf{x}$ is feasible, as required.

We define an n-vector $\mathbf{x}^0$ to be a *solution vector* for (S) if it is feasible for (S) and $\mathbf{c}^{\mathrm{T}}\mathbf{x}^0 \geq \mathbf{c}^{\mathrm{T}}\mathbf{x}$ for any feasible $\mathbf{x}$. Let X^0 denote the set of such vectors: X^0 is non-empty if and only if (S) is soluble. The set X^0 is convex. To see this, first notice that the empty set is convex. Now suppose that (S) is soluble with solution value π, that $\mathbf{x}_1$ and $\mathbf{x}_2$ are solution vectors and that $0 < \alpha < 1$. Letting $\mathbf{x} = \alpha\mathbf{x}_1 + (1-\alpha)\mathbf{x}_2$ we see that $\mathbf{x}$ is feasible (by the last paragraph) and

$$\mathbf{c}^{\mathrm{T}}\mathbf{x} = \alpha\mathbf{c}^{\mathrm{T}}\mathbf{x}_1 + (1-\alpha)\mathbf{c}^{\mathrm{T}}\mathbf{x}_2 = \alpha\pi + (1-\alpha)\pi = \pi$$

so $\mathbf{x}$ is a solution vector.

Similarly, an m-vector $\mathbf{p}$ is said to be feasible for (D) if $\mathbf{A}^{\mathrm{T}}\mathbf{p} \geq \mathbf{c}$ and $\mathbf{p} \geq \mathbf{0}$: such exists if and only if (D) is feasible. An m-vector $\mathbf{p}^0$ is said to be a solution vector for (D) if it is feasible and $\mathbf{b}^{\mathrm{T}}\mathbf{p}^0 \leq \mathbf{b}^{\mathrm{T}}\mathbf{p}$ for all feasible $\mathbf{p}$: such exists if and only if (D) is soluble. The set of feasible vectors for (D) is a convex set in R^m and the set of solution vectors is a convex subset of it.

We shall have much more to say about convexity in Chapter 10, and we shall not dwell on it here. The point to notice is that if $\mathbf{x}_1$ and $\mathbf{x}_2$ are feasible for (S) then so is $\frac{3}{5}\mathbf{x}_1 + \frac{2}{5}\mathbf{x}_2$; if $\mathbf{p}_1$ and $\mathbf{p}_2$ are solution vectors for (D) then so is $\frac{1}{4}\mathbf{p}_1 + \frac{3}{4}\mathbf{p}_2$; and so on. By a similar argument, if $\mathbf{x}_1$, $\mathbf{x}_2$ and $\mathbf{x}_3$ are feasible for (S) then so is $\frac{1}{3}(\mathbf{x}_1 + \mathbf{x}_2 + \mathbf{x}_3)$.

We now establish some relations between (S) and (D).

THEOREM 1 The dual of (D) is (S).

PROOF First notice that this *is* a theorem and not a definition: it says that if we express (D) in standard form and then take the dual, we end up with something equivalent to (S).

The standard-form version of (D) is:

$$\text{maximise } -\mathbf{b}^{\mathrm{T}}\mathbf{p} \quad \text{subject to } -\mathbf{A}^{\mathrm{T}}\mathbf{p} \leq -\mathbf{c} \text{ and } \mathbf{p} \geq \mathbf{0}.$$

Thus the dual of (D) is

$$\text{minimise } (-\mathbf{c})^{\mathrm{T}}\mathbf{x} \quad \text{subject to } (-\mathbf{A}^{\mathrm{T}})^{\mathrm{T}}\mathbf{x} \geq -\mathbf{b} \text{ and } \mathbf{x} \geq \mathbf{0}.$$

Since $(-\mathbf{A}^{\mathrm{T}})^{\mathrm{T}} = -\mathbf{A}$ we may write the dual of (D) as:

$$\text{minimise } -\mathbf{c}^{\mathrm{T}}\mathbf{x} \quad \text{subject to } -\mathbf{A}\mathbf{x} \geq -\mathbf{b} \text{ and } \mathbf{x} \geq 0.$$

The standard-form version of this minimisation programme is (S).

QED

THEOREM 2 Let $\mathbf{x}$ be feasible for (S) and $\mathbf{p}$ feasible for (D). Then $\mathbf{c}^T\mathbf{x} \le \mathbf{b}^T\mathbf{p}$.

PROOF Since $\mathbf{x} \ge \mathbf{0}$ and $\mathbf{A}^T\mathbf{p} \ge \mathbf{c}$, $\mathbf{c}^T\mathbf{x} \le (\mathbf{A}^T\mathbf{p})^T\mathbf{x}$. Since $\mathbf{Ax} \le \mathbf{b}$ and $\mathbf{p} \ge \mathbf{0}$, $\mathbf{b}^T\mathbf{p} \ge (\mathbf{Ax})^T\mathbf{p}$. But

$$(\mathbf{A}^T\mathbf{p})^T\mathbf{x} = \mathbf{p}^T\mathbf{A}\mathbf{x} = (\mathbf{A}\mathbf{x})^T\mathbf{p}$$

so $\mathbf{c}^T\mathbf{x} \le \mathbf{p}^T\mathbf{A}\mathbf{x} \le \mathbf{b}^T\mathbf{p}$. QED

Now let $\mathbf{x}^0$ be an n-vector and $\mathbf{p}^0$ an m-vector. We say that $\mathbf{x}^0$ and $\mathbf{p}^0$ satisfy the *optimality condition* for (S) and (D) if $\mathbf{x}^0$ is feasible for (S) and $\mathbf{p}^0$ is feasible for (D) and $\mathbf{c}^T\mathbf{x}^0 = \mathbf{b}^T\mathbf{p}^0$. We shall show in Section 7.1 that the optimality condition is a necessary and sufficient condition for $\mathbf{x}^0$ and $\mathbf{p}^0$ to be solution vectors for (S) and (D) respectively. For the moment, we content ourselves with establishing sufficiency.

THEOREM 3 Let $\mathbf{x}^0$ and $\mathbf{p}^0$ satisfy the optimality condition for (S) and (D). Then $\mathbf{x}^0$ and $\mathbf{p}^0$ are solution vectors for (S) and (D) respectively.

PROOF Denote the common value of $\mathbf{c}^T\mathbf{x}^0$ and $\mathbf{b}^T\mathbf{p}^0$ by α. We want to show that $\mathbf{c}^T\mathbf{x}^1 \le \alpha$ whenever $\mathbf{x}^1$ is feasible for (S), and that $\mathbf{b}^T\mathbf{p}^1 \ge \alpha$ whenever $\mathbf{p}^1$ is feasible for (D). To prove the former result, apply Theorem 2 with $\mathbf{x} = \mathbf{x}^1$ and $\mathbf{p} = \mathbf{p}^0$; to prove the latter, apply Theorem 2 with $\mathbf{x} = \mathbf{x}^0$ and $\mathbf{p} = \mathbf{p}^1$. QED

There is another, very useful, way of stating the optimality condition. This uses complementarity, as defined in Section 3.3.

THEOREM 4 $\mathbf{x}$ and $\mathbf{p}$ satisfy the optimality condition for (S) and (D) if and only if the $(n + m)$-vectors

$$\mathbf{z} = (\mathbf{x} : \mathbf{b} - \mathbf{A}\mathbf{x}) \quad \text{and} \quad \mathbf{w} = (\mathbf{A}^T\mathbf{p} - \mathbf{c} : \mathbf{p})$$

are non-negative and complementary.

PROOF A sum of non-negative terms is zero if and only if each term is zero. Thus two vectors are non-negative and complementary if and only if they are non-negative and orthogonal. We therefore want to show that $\mathbf{x}$ and $\mathbf{p}$ satisfy the optimality condition if and only if $\mathbf{z} \ge \mathbf{0}$, $\mathbf{w} \ge \mathbf{0}$ and $\mathbf{w}^T\mathbf{z} = 0$. But non-negativity of $\mathbf{z}$ is equivalent to feasibility of $\mathbf{x}$, non-negativity of $\mathbf{w}$ is equivalent to feasibility of $\mathbf{p}$ and

$$\mathbf{w}^T\mathbf{z} = \mathbf{p}^T\mathbf{A}\mathbf{x} - \mathbf{c}^T\mathbf{x} + \mathbf{p}^T\mathbf{b} - \mathbf{p}^T\mathbf{A}\mathbf{x} = \mathbf{b}^T\mathbf{p} - \mathbf{c}^T\mathbf{x}.$$ QED

Theorems 3 and 4 suggest a procedure for solving (S) and (D) simultaneously. Using pivot-and-switch, we search over complementary solutions to the pair of systems

$$\mathbf{A}\mathbf{x} + \mathbf{y} = \mathbf{b}, \quad \mathbf{A}^{\mathrm{T}}\mathbf{p} - \mathbf{q} = \mathbf{c}$$

until we find one in which $\mathbf{x}$, $\mathbf{y}$, $\mathbf{p}$, $\mathbf{q}$ are *all* non-negative: then $\mathbf{x}$ and $\mathbf{p}$ are solution vectors for (S) and (D) respectively.

Before we can make such a procedure operational we have to surmount two difficulties. The first is that of finding a method of choosing our pivot entries. The second is more conceptual. Clearly a 'method of solving (S) and (D)' can only work when both these programmes are in fact soluble. What we really need is a procedure which not only solves (S) and (D) when both are soluble, but also flashes the appropriate warning-lights when at least one of them is not.

The whole of Chapter 6 is addressed to these issues.

Exercises

1. Formulate the duals of Examples 5.1.1 and 5.1.2.
2. Show that if (S) is feasible and unbounded then (D) is infeasible.

6 The Simplex Method

Let the matrix **A**, the vectors **b** and **c** and the linear programmes (S) and (D) be as in Chapter 5. We are interested in complementary solutions to the pair of systems

$$\mathbf{A}\mathbf{x} + \mathbf{y} = \mathbf{b}, \quad \mathbf{A}^{\mathrm{T}}\mathbf{p} - \mathbf{q} = \mathbf{c}. \tag{1}$$

If we find one for which

$$\mathbf{x} \geq \mathbf{0}, \quad \mathbf{y} \geq \mathbf{0}, \quad \mathbf{p} \geq \mathbf{0}, \quad \mathbf{q} \geq \mathbf{0} \tag{2}$$

then **x** and **p** are solution vectors for (S) and (D) respectively.

Let $\mathbf{z} = (\mathbf{x} : \mathbf{y})$, $\mathbf{w} = (\mathbf{q} : \mathbf{p})$. Let

$$\Sigma = \begin{array}{c|c|ccc} & & \pi_1 & \cdots & \pi_n \\ \hline & \alpha_{00} & \alpha_{01} & \cdots & \alpha_{0n} \\ \hline \pi_{n+1} & \alpha_{10} & \alpha_{11} & & \alpha_{1n} \\ \vdots & \vdots & & \vdots & \\ \pi_{n+m} & \alpha_{m0} & \alpha_{m1} & & \alpha_{mn} \end{array}$$

be a scheme for (1), obtained by pivot-and-switch. Given Σ, we can read off a complementary solution $\bar{\mathbf{z}} = (\bar{\mathbf{x}} : \bar{\mathbf{y}})$, $\bar{\mathbf{w}} = (\bar{\mathbf{q}} : \bar{\mathbf{p}})$ as follows:

$$\bar{z}_k = \begin{cases} \alpha_{i0} & \text{if } k = \pi_{n+i} \text{ for some } i > 0 \\ 0 & \text{if } k \text{ labels a column} \end{cases}$$

$$\bar{w}_k = \begin{cases} \alpha_{0j} & \text{if } k = \pi_j \text{ for some } j \leq n \\ 0 & \text{if } k \text{ labels a row.} \end{cases}$$

Recall from Section 3.3 that $\mathbf{c}^{\mathrm{T}}\bar{\mathbf{x}} = \mathbf{b}^{\mathrm{T}}\bar{\mathbf{p}} = \alpha_{00}$.

Clearly $\bar{\mathbf{x}}$ is feasible for (S) if and only if $\bar{\mathbf{z}} \geq \mathbf{0}$, which is the same as saying that

$$\alpha_{i0} \geq 0 \quad \text{for } i = 1, \ldots, m.$$

In this case Σ is said to be a *row feasible* scheme.

Similarly $\bar{\mathbf{p}}$ is feasible for (D) if and only if $\bar{\mathbf{w}} \geq \mathbf{0}$, which is the same as saying that

$$\alpha_{0j} \geq 0 \quad \text{for } j = 1, \ldots, n.$$

In this case Σ is said to be *column feasible*.

If Σ is both row and column feasible then $\bar{\mathbf{x}}, \bar{\mathbf{y}}, \bar{\mathbf{p}}, \bar{\mathbf{q}}$ satisfy (2) so $\bar{\mathbf{x}}$ is a solution vector for (S) and $\bar{\mathbf{p}}$ for (D). In this case α_{00} is the solution value for *both* programmes. But how do we obtain such a scheme?

6.1 The pivot step

This section is concerned with the question: given a row feasible scheme, what do we do with it? Our remarks about $\bar{\mathbf{x}}$ show that a necessary condition for a row feasible scheme to exist is that (S) be feasible: it will be shown in Section 6.3 that this condition is also sufficient. These issues of existence of row feasible schemes will be dealt with in good time, and the question with which we began is enough to keep us going for one section. So we assume for the rest of this section that the given scheme Σ is row feasible.

To illustrate our analysis, we shall use some mnemonic diagrams of schemes. In each diagram,

$\oplus$ will denote a non-negative number
$\ominus$ will denote a non-positive number
$+$ will denote a positive number

and $-$ will denote a negative number.

Thus our row-feasibility assumption says that Σ is of the form

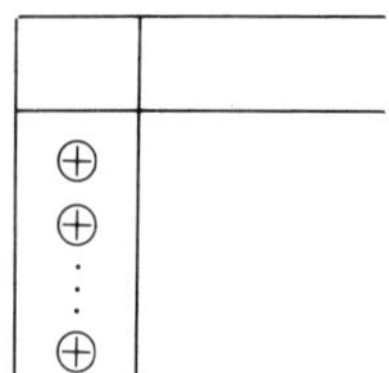

Given that Σ is row feasible, one and only one of the following three cases can arise.

Case 1 $\alpha_{0j} \geq 0$ for $j = 1, \ldots, n$.
Case 2 There exists some h, with $1 \leq h \leq n$, such that $\alpha_{0h} < 0$ and $\alpha_{ih} \leq 0$ for $i = 1, \ldots, m$.
Case 3 Neither Case 1 nor Case 2 obtains.

Cases 1 and 2 are illustrated below. Case 3 occurs when some labelled column has a negative top entry and each such column has at least one positive entry.

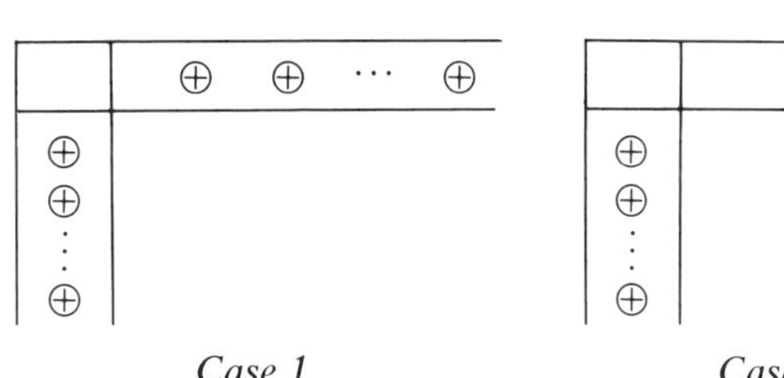

Case 1 *Case 2*

We now consider each case in turn.

Case 1
In this case Σ is both row feasible and column feasible, so we can read off solution vectors for (S) and (D) and the common solution value α_{00}.

Case 2
Here we may choose an integer h such that $1 \leq h \leq n$, $\alpha_{0h} < 0$ and $\alpha_{ih} \leq 0$ for $i = 1, \ldots, m$. We shall show that in this case (S) *is feasible and unbounded and* (D) *is infeasible.*

To prove the result about (S) we consider the row system of Σ. Given an $(n + m)$-vector $\mathbf{z} = (\mathbf{x} : \mathbf{y})$ we have $\mathbf{Ax} + \mathbf{y} = \mathbf{b}$ if and only if

$$z_{\pi_{n+i}} + \sum_{j=1}^{n} \alpha_{ij} z_{\pi_j} = \alpha_{i0} \quad \text{for } i = 1, \ldots, m, \tag{1}$$

in which case

$$\mathbf{c}^{\mathrm{T}}\mathbf{x} = \alpha_{00} - \sum_{j=1}^{n} \alpha_{0j} z_{\pi_j}. \tag{2}$$

Let M be any positive number and define the $(n + m)$-vector $\mathbf{z}(M) = (\mathbf{x}(M) : \mathbf{y}(M))$ by setting

$$\begin{aligned} &z_k(M) = M \quad \text{if } k = \pi_h \\ &z_k(M) = \alpha_{i0} - M\alpha_{ih} \quad \text{if } k = \pi_{n+i} \quad (i = 1, \ldots, m) \\ &z_k(M) = 0 \quad \text{otherwise.} \end{aligned}$$

It is easy to see that $\mathbf{z}(M)$ satisfies (1): all terms in the sum vanish except for $j = h$. By row-feasibility of Σ and our choice of h, $\mathbf{z}(M) \geq 0$. Thus

$$\mathbf{Ax}(M) + \mathbf{y}(M) = \mathbf{b},\ \mathbf{x}(M) \geq \mathbf{0},\ \mathbf{y}(M) \geq \mathbf{0}$$

so that $\mathbf{x}(M)$ is feasible for (S). But by (2),

$$\mathbf{c}^{\mathrm{T}}\mathbf{x}(M) = \alpha_{00} - M\alpha_{0h}.$$

Since $\alpha_{0h} < 0$, we can make $\mathbf{c}^{\mathrm{T}}\mathbf{x}(M)$ as large as we like by choosing M large enough. Thus (S) is feasible and unbounded.

To prove that (D) is infeasible, we could appeal to the result of Exercise 5.3.2. Alternatively, we can work directly with the column system of Σ. Let $\mathbf{w} = (\mathbf{q} : \mathbf{p})$ be any $(n + m)$-vector, and let

$$f_j(\mathbf{w}) = w_{\pi_j} - \sum_{i=1}^{m} \alpha_{ij} w_{\pi_{n+i}}$$

for $j = 1, \ldots, n$. Then $\mathbf{A}^{\mathrm{T}}\mathbf{p} - \mathbf{q} = \mathbf{c}$ if and only if

$$f_j(\mathbf{w}) = \alpha_{0j} \quad \text{for } j = 1, \ldots, n.$$

By our choice of h, $f_h(\mathbf{w})$ is a linear form in $\mathbf{w}$ with non-negative coefficients, and $\alpha_{0h} < 0$. Thus

$$\text{if } \mathbf{w} \geq \mathbf{0}, \text{ then } f_h(\mathbf{w}) \geq 0 > \alpha_{0h}.$$

We have now proved that the equation $\mathbf{A}^{\mathrm{T}}\mathbf{p} - \mathbf{q} = \mathbf{c}$ is incompatible with the inequalities $\mathbf{p} \geq \mathbf{0}$, $\mathbf{q} \geq \mathbf{0}$, so (D) is infeasible.

Case 3

Here neither Case 1 nor Case 2 obtains. Since we are not in Case 1, we can choose some integer s such that $1 \leq s \leq n$ and $\alpha_{0s} < 0$. Since we are not in Case 2, $\alpha_{is} > 0$ for some $i > 1$. So the situation may be depicted as follows.

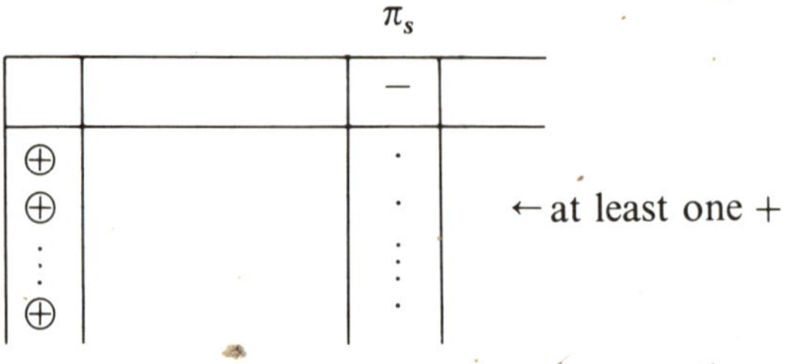

		π_s		
		−		
⊕		·		
⊕		·		← at least one +
⋮		⋮		
⊕		·		

We may now choose an integer r such that

$i = r$ minimises the ratio α_{i0}/α_{is} among all those i such that $1 \leq i \leq m$ and $\alpha_{is} > 0$. (3)

Suppose we *pivot and switch using* α_{rs} *as pivot entry*. This gives a new scheme, say

$$\Sigma^* = \begin{array}{c|c|ccc} & & \psi_1 & \cdots & \psi_n \\ \hline & \beta_{00} & \beta_{01} & \cdots & \beta_{0n} \\ \hline \psi_{n+1} & \beta_{10} & \beta_{11} & \cdots & \beta_{1n} \\ \vdots & \vdots & & \vdots & \\ \psi_{n+m} & \beta_{m0} & \beta_{m1} & \cdots & \beta_{mn} \end{array}$$

We now show that

Σ^* is row-feasible, $\beta_{00} \geq \alpha_{00}$ and $\beta_{0s} > 0$. (4)

Let $\mu = \alpha_{r0}/\alpha_{rs}$. Then $\mu \geq 0$. Let i be any integer such that $1 \leq i \leq m$. If $i = r$, then $\beta_{i0} = \mu \geq 0$. If $i \neq r$ and $\alpha_{is} > 0$, then $\alpha_{i0}/\alpha_{is} \geq \mu$ by (3), so

$$\beta_{i0} = \alpha_{i0} - \mu\alpha_{is} = \alpha_{is}((\alpha_{i0}/\alpha_{is}) - \mu) \geq 0.$$

If $i \neq r$ and $\alpha_{is} \leq 0$ then $\mu\alpha_{is} \leq 0 \leq \alpha_{i0}$ so $\beta_{i0} = \alpha_{i0} - \mu\alpha_{is} \geq 0$.

Thus $\beta_{i0} \geq 0$ for all $i = 1, \ldots, m$ so Σ^* is row feasible. Also $\alpha_{0s} < 0$ by our choice of s, so

$$\beta_{00} - \alpha_{00} = -\mu\alpha_{0s} \geq 0.$$

Finally, $\beta_{0s} = -\alpha_{0s}/\alpha_{is} > 0$.

The row feasibility of Σ^* means that we can classify it into Case 1, 2, or 3 and repeat the procedure. Notice that (4) gives us two justifications for the rule (3) for choosing the pivot entry. One, based on solving (S), appeals to $\beta_{00} \geq \alpha_{00}$: the new feasible $\mathbf{z}$ read off from Σ^* gives a value of $\mathbf{c}^{\mathrm{T}}\mathbf{x}$ at least as large as the old one. The other, based on the search for non-negative complementarity, appeals to

$$\beta_{0s} > 0 > \alpha_{0s}.$$

We replace a negative entry on top of the column s by a positive one.

This case-by-case analysis tells us what to do, given a row feasible scheme Σ.

In Case 1, we have row and column feasibility and can read off solution vectors and the common solution value for (S) and (D).

In Case 2, we observe that (S) is feasible and unbounded and proceed no further.

In Case 3, we choose some $s \geq 1$ such that $\alpha_{0s} < 0$. We then choose an integer r such that

> $i = r$ minimises the ratio α_{i0}/α_{is} among all those i such that $1 \leq i \leq m$ and $\alpha_{is} > 0$.

We then pivot-and-switch with α_{rs} as pivot entry. This yields a new row-feasible scheme whose top-left-corner entry is not less than α_{00}. This new scheme may be classified as Case 1, 2 or 3. We repeat the procedure until Case 1 or Case 2 is reached.

This iterative procedure is known as the *simplex method*. We discuss it in more detail in the next section.

6.2 Implementation

The first item on the agenda is: *where do we begin?*

It is clear from our analysis of Case 3 that once an initial row-feasible scheme has been chosen, all subsequent schemes generated by the method will be row-feasible. But how do we choose our initial row-feasible scheme?

There is one important special case where the initialisation problem is trivial, namely $\mathbf{b} \geq 0$. For in that case, the scheme

$$\Sigma^0 = \begin{array}{c|c|c} & & 1 \quad \cdots \quad n \\ \hline & 0 & -\mathbf{c}^{\mathrm{T}} \\ \hline n+1 & & \\ \vdots & \mathbf{b} & \mathbf{A} \\ n+m & & \end{array}$$

is row-feasible. Since Σ^0 can be written down by inspection of $\mathbf{A}$, $\mathbf{b}$ and $\mathbf{c}$, it is the natural initial scheme when $\mathbf{b} \geq \mathbf{0}$. If some component of $\mathbf{b}$ is negative, Σ^0 is not row-feasible and we have to work harder: indeed in this case it may turn out that no row-feasible scheme exists. The problems arising from negative components of $\mathbf{b}$ will be dealt with in Section 6.3 where it is shown that a row-feasible scheme exists if and only if (S) is feasible. The example and exercises of this section will be restricted to the case where $\mathbf{b} \geq \mathbf{0}$.

But before we attack these examples, there are three loose ends to be tied up.

(i) Suppose we are in Case 3 and there is more than one admissible pivot column (that is, more than one s for which $\alpha_{0s} < 0$). Which one do we choose for our pivot column?

(ii) Suppose we are in Case 3 and have chosen the pivot column s, and that there is more than one admissible pivot row. This case arises when, for example $\alpha_{is} > 0$ for $i = 1$, 4 and 8 and

$$\alpha_{10}/\alpha_{1s} = \alpha_{80}/\alpha_{8s} < \alpha_{40}/\alpha_{4s}.$$

Here $r = 1$ and $r = 8$ both satisfy the conditions for an admissible pivot row: which should we choose?

(iii) Can we be sure of reaching Case 1 or Case 2 after a finite number of pivots?

One possible answer to both (i) and (ii) is: *when in doubt, choose the column or row with the smallest label.* This rule is due to Bland (1977) and we shall refer to it as *Bland's Refinement* of the simplex method. Its purpose is to provide a positive answer to question (iii), to which we now turn.

It is in the nature of the simplex method that one is always moved on from a Case 3 scheme via a pivot-and-switch operation: thus no scheme can be identical to its immediate predecessor. Also, since **A** has only a finite number of rows and columns, there are only finitely many schemes for our pair of equation systems. These facts imply that *if* each scheme generated by the simplex method differs from *all* its predecessors, then Case 1 or Case 2 must be reached after a finite number of pivots. Thus the only situation where termination is *not* achieved after a finite number of operations is typified by the case where the 9th scheme is the same as the 6th, the 10th scheme the same as the 7th, the 11th the same as the 8th, the 12th the same as the 9th and 6th and so on. This sad state of affairs is known as *cycling* (or *circling*).

To explore this in more detail, suppose that we start from the scheme Σ^0 and that subsequent schemes generated by the simplex method are $\Sigma^1, \Sigma^2, \ldots$. For $t = 0, 1, 2, \ldots$ denote the top-left-corner (or α_{00}) entry of Σ^t by λ_t. By our analysis of Case 3,

$$\lambda_0 \leq \lambda_1 \leq \lambda_2 \leq \cdots.$$

Also notice that if the left-border entry in the pivot row of Σ^t is not zero, then $\lambda_{t+1} > \lambda_t$, so Σ^{t+1} differs from all earlier schemes. All this suggests that the simplex method has some inbuilt resistance to cycling. But unfortunately this resistance is not total: examples have been constructed to show that the ordinary simplex method can cycle.

This is where Bland's Refinement comes in. In Appendix A we shall give a proof, adapted from Bland's article, that when the simplex method is applied *with Bland's Refinement*, cycling cannot occur. It should be noted that the world did not have to wait until 1977 for a procedure that would avoid cycling in the simplex method: but earlier procedures, the most famous being the 'lexicographic simplex method', were considerably more difficult to remember and to apply than Bland's Refinement.

We can now state and prove the fundamental theorem on the simplex method.

SIMPLEX THEOREM If there exists a row feasible scheme then
either (S) and (D) are both soluble, with the same solution value
or (S) is feasible and unbounded and (D) is infeasible.

PROOF Starting from any row feasible scheme we apply the simplex method, using Bland's Refinement if necessary to avoid cycling. After a finite number of pivot steps we obtain Case 1 (*either*) or Case 2 (*or*).

QED

Our purpose in discussing cycling and Bland's Refinement was to obtain a rigorous proof of this theorem: Appendix A completes the proof. Having said all this, we shall typically not use Bland's Refinement in examples. We adopt instead the following *ad hoc* rules for answering questions (i) and (ii):

(i) when choosing between admissible pivot columns, choose one for which $(-\alpha_{0s})$ is largest;
(ii) when choosing between admissible pivot rows, choose at random or in the interests of computational simplicity.

These *ad hoc* rules, while they usually lead to rather quicker solutions than Bland's Refinement, can give rise to cycling. However, *experience has shown that cycling is a very rare phenomenon*, so the importance of anti-cycling procedures is largely theoretical rather than computational. (There are ways of splicing together Bland's Refinement and the *ad hoc* rules which have the advantages of both. Using 'maximise $-\alpha_{0s}$' for columns and 'smallest label' for rows is not one of them: that can cycle.)

Example 1 We solve the 'widget' problem of Example 5.1.2. This is an example of (S) with

$$\mathbf{A} = \begin{pmatrix} 1 & 0 & 0 \\ 1 & 2 & 0 \\ 0 & 3 & 4 \\ -1 & -1 & 1 \end{pmatrix}, \quad \mathbf{b} = \begin{pmatrix} 20 \\ 50 \\ 65 \\ 0 \end{pmatrix}, \quad \mathbf{c} = \begin{pmatrix} 120 \\ 100 \\ 8 \end{pmatrix}.$$

Since $\mathbf{b} \geq \mathbf{0}$ we can start with the scheme

		1	2	3
	0	−120	−100	−8
4	20	(1)	0	0
5	50	1	2	0
6	65	0	3	4
7	0	−1	−1	1

This is a Case 3 scheme. Since 120 is the largest of the numbers 120, 100, 8 we choose our pivot entry in the column labelled '1'. Since $\frac{20}{1}$ is the smaller of the numbers $\frac{20}{1}$ and $\frac{50}{1}$ we choose our pivot entry in the row labelled '4'. Pivoting as shown, we obtain the scheme

		4	2	3
	2400	120	−100	−8
1	20	1	0	0
5	30	−1	(2)	0
6	65	0	3	4
7	20	1	−1	1

We are still in Case 3. Since $100 > 8$, we pivot in the column labelled '2'. Since $\frac{30}{2} < \frac{65}{3}$, we pivot as shown. This gives the scheme

		4	5	3
	3900	70	50	−8
1	20	1	0	0
2	15	$-\frac{1}{2}$	$\frac{1}{2}$	0
6	20	$\frac{3}{2}$	$-\frac{3}{2}$	(4)
7	35	$\frac{1}{2}$	$\frac{1}{2}$	1

Again we are in Case 3. The only admissible pivot column is

labelled '3'. Since $\frac{20}{4} < \frac{35}{1}$ we pivot as shown to get the scheme

		4	5	6
	3940	73	47	2
1	20			
2	15			
3	5			
7	30			

We are now in Case 1 and have finished (which is why we have not bothered to calculate the entries in the last scheme, apart from those on the borders). A solution to the programme is

$$x_1 = 20, \quad x_2 = 15, \quad x_3 = 5$$

and the solution value is 3940.

Referring back to Example 5.1.2 we see that the firm should produce 35 widgets per week (20 by technique A and 15 by technique B) and turn 5 of them into superwidgets. Profit is £3940 per week.

Example 2

$$\begin{aligned} &\text{Maximise} && -15x_1 + 6x_2 + 4x_3 \\ &\text{subject to} && 2x_1 - x_2 - x_3 \le 1 \\ &&& -8x_1 + 2x_2 + x_3 \le 2 \\ &&& x_1 \ge 0, x_2 \ge 0, x_3 \ge 0. \end{aligned}$$

Our initial scheme is

		1	2	3
	0	15	-6	-4
4	1	2	-1	-1
5	2	-8	(2)	1

Since $6 > 4$ we choose our pivot entry in the column labelled '2'. The only positive entry in that column is in the row labelled '5' so we pivot there. This gives us the scheme

		1	5	3
	6	-9	3	-1
4	2	-2	$\frac{1}{2}$	$-\frac{1}{2}$
2	1	-4	$\frac{1}{2}$	$\frac{1}{2}$

In this scheme, all entries in the column labelled '1' are negative. We are therefore in Case 2, so that our programme is *feasible and unbounded*.

To check that this is so, we repeat the analysis of Case 2 given in the last section. Let M be any positive number and let

$$x_1 = M, \quad x_2 = 1 + 4M, \quad x_3 = 0.$$

Then all constraints are satisfied and the objective function takes the value $6 + 9M$, which becomes indefinitely large as $M \to \infty$.

Neither of the last two examples exploits the ability of the simplex method to handle (S) and (D) simultaneously. Our next example does this.

Example 3 We solve the diet problem of Example 5.1.1. The problem is to

$$\text{minimise } 25p_1 + 40p_2 \quad \text{subject to}$$
$$3p_1 + p_2 \geq 8, \quad 4p_1 + 3p_2 \geq 19, \quad p_1 + 2p_2 \geq 7, \; p_1 \geq 0, \quad p_2 \geq 0.$$

This is an example of (D) with

$$\mathbf{A} = \begin{pmatrix} 3 & 4 & 1 \\ 1 & 3 & 2 \end{pmatrix}, \quad \mathbf{b} = \begin{pmatrix} 25 \\ 40 \end{pmatrix}, \quad \mathbf{c} = \begin{pmatrix} 8 \\ 19 \\ 7 \end{pmatrix}.$$

The simplex method takes us through the following schemes:

		1	2	3
	0	-8	-19	-7
4	25	3	(4)	1
5	40	1	3	2

		1	4	3
	$\frac{475}{4}$	$\frac{25}{4}$	$\frac{19}{4}$	$-\frac{9}{4}$
2	$\frac{25}{4}$	$\frac{3}{4}$	$\frac{1}{4}$	$\frac{1}{4}$
5	$\frac{85}{4}$	$-\frac{5}{4}$	$-\frac{3}{4}$	($\frac{5}{4}$)

		1	4	5
	157	4	$\frac{17}{5}$	$\frac{9}{5}$
2	2			
3	17			

From the top row of the last (Case 1) scheme we see that a solution to (D) is $p_1 = \frac{17}{5}$, $p_2 = \frac{9}{5}$ and that the solution value is 157.

Thus the cost-minimising diet consists of 3.4 units of bread and 1.8 units of cheese and costs 157 units of money.

Exercises

1. Maximise $x_1 + 2x_2 + x_3$ subject to
$$\begin{aligned} 2x_1 + x_2 - x_3 &\leq 2 \\ 3x_1 - x_2 + 5x_3 &\leq 7 \\ 4x_1 + x_2 + x_3 &\leq 6 \\ x_1 \geq 0, x_2 \geq 0, x_3 &\geq 0. \end{aligned}$$

2. Solve the profit-maximisation problem of Exercise 5.1.1.

3. Minimise $3x_1 - x_2 + x_3 + x_4$ subject to
$$\begin{aligned} x_1 - x_2 + 2x_3 - x_4 &\leq 1 \\ 2x_1 + 4x_2 - 5x_3 + x_4 &\leq 2 \\ x_1 \geq 0, x_2 \geq 0, x_3 \geq 0, x_4 &\geq 0. \end{aligned}$$

4. Minimise $3p_1 + 2p_2 + p_3 + 4p_4$ subject to
$$\begin{aligned} 2p_1 + 4p_2 + 5p_3 + p_4 &\geq 10 \\ 3p_1 - p_2 + 7p_3 - 2p_4 &\geq 2 \\ 5p_1 + 2p_2 + p_3 + 6p_4 &\geq 5 \\ p_1 \geq 0, p_2 \geq 0, p_3 \geq 0, p_4 &\geq 0. \end{aligned}$$

5. Minimise $2p_1 + 2p_2 + 4p_3$ subject to
$$\begin{aligned} 4p_1 + 3p_2 + 5p_3 &\geq 2 \\ 3p_1 + p_2 + 7p_3 &\leq 1 \\ p_1 + 4p_2 + 6p_3 &\leq 3 \\ p_1 \geq 0, p_2 \geq 0, p_3 &\geq 0. \end{aligned}$$

6. Maximise $x_1 + x_2 - x_3$ subject to
$$\begin{aligned} 6x_1 - 2x_2 - x_3 &\leq 4 \\ 2x_1 + 7x_2 + 2x_3 &\leq 1 \\ x_1 \geq 0, x_2 \geq 0, x_3 &\text{ free.} \end{aligned}$$

7. Show that the linear programme

maximise $x_1 + x_2 + x_3$ subject to
$$\begin{aligned} x_1 - x_2 + x_3 &\leq 1 \\ x_1 + x_2 - x_3 &\leq 2 \\ 4x_1 + 2x_2 - 3x_3 &\leq 0 \\ x_1 \geq 0, x_2 \geq 0, x_3 &\geq 0 \end{aligned}$$

is feasible and unbounded.

Hence or otherwise show that if

$$\begin{aligned} p_1 + p_2 + 4p_3 &\geq 1 \\ -p_1 + p_2 + 2p_3 &\geq 1 \\ p_1 - p_2 - 3p_3 &\geq 1 \end{aligned}$$

then at least one of p_1, p_2, p_3 is negative.

6.3 Auxiliary programmes

As in earlier sections, let $\mathbf{A}$ be an $m \times n$ matrix, $\mathbf{b}$ an m-vector and $\mathbf{c}$ an n-vector. Let the standard-form programme (S) and its dual (D) be formulated in the usual way. To apply the simplex method, we need an initial row-feasible scheme. If $\mathbf{b} \geq \mathbf{0}$, such a scheme can be written down immediately. In this section we consider the case where at least one component of $\mathbf{b}$ is negative.

The first thing to notice about this case is that (S) may or may not be feasible. If, for example,

$$\mathbf{A} = \begin{pmatrix} 1 & 1 \\ -2 & 1 \end{pmatrix} \quad \text{and} \quad \mathbf{b} = \begin{pmatrix} 1 \\ -1 \end{pmatrix},$$

then $\mathbf{x} = (1 \quad 0)^{\mathrm{T}}$ is feasible for (S). On the other hand, if

$$\mathbf{A} = \begin{pmatrix} 1 & 1 \\ 2 & 1 \end{pmatrix} \quad \text{and} \quad \mathbf{b} = \begin{pmatrix} 1 \\ -1 \end{pmatrix},$$

then (S) is infeasible. In general, we want to be able to *test* whether or not (S) is feasible.

To this end we consider the following standard-form programme in $(1 + n)$ variables $x_*, x_1, \ldots, x_n$:

$$\begin{aligned} &\text{maximise} && -x_* \\ &\text{subject to} && -x_* + a_{11}x_1 + \cdots + a_{1n}x_n \leq b_1 \\ &&& \qquad\vdots \\ &&& -x_* + a_{m1}x_1 + \cdots + a_{mn}x_n \leq b_m \\ &&& x_* \geq 0,\ x_1 \geq 0,\ \ldots,\ x_n \geq 0. \end{aligned}$$

This is called the *auxiliary programme* and may be written in matrix form as

$$\text{maximise } -x_* \text{ subject to } (-\mathbf{e}, \mathbf{A})\begin{pmatrix} x_* \\ \mathbf{x} \end{pmatrix} \leq \mathbf{b} \text{ and } \begin{pmatrix} x_* \\ \mathbf{x} \end{pmatrix} \geq \mathbf{0}$$

where $\mathbf{x} = (x_1 \ \ldots \ x_n)^{\mathrm{T}}$ and $\mathbf{e}$ is the m-vector consisting entirely of 1's.

The auxiliary programme is feasible, since the constraints may be satisfied by choosing $\mathbf{x} = \mathbf{0}$ and x_* sufficiently large that $x_* \geq -b_i$ for $i = 1, \ldots, m$. If the $(1 + n)$-vector $(x_* : \mathbf{x})$ is feasible for the auxiliary programme, then $-x_* \leq 0$; further, $(0 : \mathbf{x})$ is feasible for the auxiliary programme if and only if $\mathbf{x}$ is feasible for (S). This tells us two things: first, that the auxiliary programme is always feasible and bounded; second, that (S) is feasible if and only if the auxiliary programme is soluble with solution value zero.

Thus to test (S) for feasibility, it is sufficient to solve the auxiliary programme and see if the solution value is zero. This test is easy to perform since, as will shortly be demonstrated, a row-feasible scheme for the auxiliary programme may be obtained by one PS operation; we can then proceed by the simplex method for the auxiliary programme. It is computationally convenient to work with two rows above the line, with the second row corresponding to the objective function for the auxiliary programme and the top row to the objective function for the original programme (S). If we find that (S) is feasible we can pass straight to a row-feasible scheme for (S).

We now go into the details. To solve the auxiliary programme, we start with the scheme

		$*$	1	$\cdots$	n
	0	0	$-c_1$	$\cdots$	$-c_n$
	0	1	0	$\cdots$	0
$n+1$	b_1	-1	a_{11}	$\cdots$	a_{1n}
$\vdots$	$\vdots$	$\vdots$		$\vdots$	
$n+m$	b_m	-1	a_{m1}	$\cdots$	a_{mn}

where the top row refers to the objective function for (S) and the row immediately above the line refers to the objective function for the auxiliary programme. Notice that this scheme is *not* row feasible, even for the auxiliary programme, since $b_i < 0$ for some i. But row-feasibility can be achieved in one step by pivoting on the '-1' in the column labelled $*$ and the row labelled '$n + l$', where b_l is the 'most negative' of the components of $\mathbf{b}$ ($b_l \leq b_i$ for all i).

We can then proceed to solve the auxiliary programme by the simplex method. The auxiliary programme is feasible and bounded, with an objective function that cannot be positive. Thus we end with a scheme of the form

		π_*	π_1	$\cdots$	π_n
	α_{00}	α_{0*}	α_{01}	$\cdots$	α_{0n}
	α_{*0}	α_{**}	α_{*1}	$\cdots$	α_{*n}
π_{n+1}	α_{10}	α_{1*}	α_{11}	$\cdots$	α_{1n}
$\vdots$	$\vdots$			$\vdots$	
π_{n+m}	α_{m0}	α_{m*}	α_{m1}	$\cdots$	α_{mn}

where

$$\alpha_{i0} \geq 0 \text{ for } i = 1, \ldots, m,$$
$$\alpha_{*j} \geq 0 \text{ for } j = 1, \ldots, n,$$
$$\alpha_{*0} \leq 0,$$

and $(\pi_*, \pi_1, \ldots, \pi_{n+m})$ is a permutation of $(*, 1, \ldots, n+m)$.

There are now three possible cases:

(i) $\alpha_{*0} < 0$;

(ii) $\alpha_{*0} = 0$ and $*$ labels a column;

(iii) $\alpha_{*0} = 0$ and $*$ labels a row.

In case (i), the auxiliary programme has a negative solution value, so (S) is infeasible.

In case (ii), we strike out the row immediately above the line and the column labelled $*$. This leaves us with a row-feasible scheme for (S). We can then proceed with the simplex method on (S).

Case (iii) rarely occurs in practice, but requires discussion, for the sake of completeness. Suppose that $* = \pi_{n+r}$ for some $r \geq 1$, and that $\alpha_{*0} = 0$. Since the objective function for the auxiliary programme is $-x_*$, $\alpha_{r0} = -\alpha_{*0} = 0$. Since x_* is not *identically* zero in the auxiliary programme, we can choose some $s \geq 1$ such that $\alpha_{rs} \neq 0$. Now pivot-and-switch with α_{rs} as pivot entry: since $\alpha_{r0} = 0$, this PS operation preserves row-feasibility whether α_{rs} is positive or negative. We then have a scheme in which the $(*, 0)$ entry is again zero and $*$ labels a column, and proceed as in case (ii).

Example

$$\begin{array}{ll} \text{Maximise} & 4x_1 - 6x_2 - 16x_3 + x_4 \\ \text{subject to} & x_1 - x_2 - x_3 - 2x_4 \leq -1 \\ & -x_1 + 3x_2 + 4x_3 + x_4 \leq -2 \\ & x_1 - 3x_2 - 5x_3 + x_4 \leq 12 \\ & x_1 \geq 0, x_2 \geq 0, x_3 \geq 0, x_4 \geq 0. \end{array}$$

To test for feasibility, we start with the scheme

		*	1	2	3	4
	0	0	−4	6	16	−1
	0	1	0	0	0	0
5	−1	−1	1	−1	−1	−2
6	−2	**−1**	−1	3	4	1
7	12	−1	1	−3	−5	1

Since $-2 < -1 < 12$, we pivot as shown, obtaining the scheme

		6	1	2	3	4
	0	0	−4	6	16	−1
	−2	1	−1	3	4	1
5	1	−1	**2**	−4	−5	−3
*	2	−1	1	−3	−4	−1
7	14	−1	2	−6	−9	0

We may now proceed with the simplex method for the auxiliary programme.

		6	5	2	3	4
	2	−2	2	2	6	7
	$-\frac{3}{2}$	$\frac{1}{2}$	$\frac{1}{2}$	1	$\frac{3}{2}$	$-\frac{1}{2}$
1	$\frac{1}{2}$	$-\frac{1}{2}$	$\frac{1}{2}$	−2	$-\frac{5}{2}$	$-\frac{3}{2}$
*	$\frac{3}{2}$	$-\frac{1}{2}$	$-\frac{1}{2}$	−1	$-\frac{3}{2}$	$\frac{1}{2}$
7	13	0	−1	−2	−4	3

		6	5	2	3	*
	23	−9	−5	−16	−15	14
	0	0	0	0	0	1
1	5	−2	−1	−5	−7	3
4	3	−1	−1	−2	−3	2
7	4	3	2	4	5	−6

We can now delete the column labelled * and the row corresponding to the auxiliary-programme objective function, thereby obtaining a row-feasible scheme for the original programme.

		6	5	2	3
	23	−9	−5	−16	−15
1	5	−2	−1	−5	−7
4	3	−1	−1	−2	−3
7	4	3	2	(4)	5

Solution of our problem now requires just one PS operation.

		6	5	7	3
	39	3	3	4	5
1	10				
4	5				
2	1				

Thus a solution is

$$x_1 = 10, \quad x_2 = 1, \quad x_3 = 0, \quad x_4 = 5$$

and the solution value is 39.

We conclude this section by summarising our progress on the theoretical front.

FEASIBILITY THEOREM If (S) is feasible a row-feasible scheme exists.
PROOF If $\mathbf{b} \geq \mathbf{0}$, the usual initial scheme is row-feasible. If $\mathbf{b}$ has at least one negative component we can use the auxiliary programme method, employing Bland's Refinement if necessary to avoid cycling. Given that (S) is feasible, we obtain a row-feasible scheme for (S) after a finite number of auxiliary-programme pivots. QED

Exercise

$$\begin{array}{ll}
\text{Maximise} & 2x_1 + x_2 + x_3 \\
\text{subject to} & x_1 - x_2 - x_3 \leq -1 \\
& 2x_1 + 3x_2 + 4x_3 \leq 5 \\
& x_1 - 3x_2 + x_3 \leq -2 \\
& x_1 \geq 0, x_2 \geq 0, x_3 \geq 0.
\end{array}$$

7 Duality in Linear Programming

All the computation of the last chapter has produced two very important theorems: the Simplex Theorem of Section 6.2 and the Feasibility Theorem of Section 6.3. In the first section of this chapter we use these results to derive existence theorems on solutions to standard-form linear programmes and their duals.

When we discussed existence of solutions to linear equation systems in Chapter 2, we also discussed uniqueness. A useful theorem on uniqueness of solutions to linear programmes is stated and proved in Section 7.2.

The pivot-and-switch formulation of the simplex method is probably the neatest way to solve linear programmes by hand, but one can instruct a computer to do better. This improvement, known as the Revised Simplex Method, is also discussed in Section 7.2.

Section 7.3 gives the 'marginalist' interpretation of solutions to the duals of production planning problems. This is a key result in mathematical economics. Section 7.4 uses this interpretation of the dual to tie up some loose ends concerning complementarity.

Throughout this chapter, (S) and (D) have the same meanings as in Chapters 5 and 6.

7.1 The Duality Theorem

We begin by stating and proving the main result in the theory of linear programming.

THEOREM 1 (Duality Theorem)

(i) If (S) is feasible and bounded, then (S) and (D) are both soluble, with the *same* solution value.

(ii) If (S) is feasible and unbounded, then (D) is infeasible.

(iii) If (S) is infeasible, then (D) is either infeasible or feasible-and-unbounded.

PROOF First suppose that (S) is feasible. By the Feasibility Theorem there exists a row-feasible scheme for (S). Now apply the Simplex Theorem: since (S) is either bounded or unbounded, but not both, we obtain (i) and (ii).

To prove (iii), it suffices to show that if (D) is feasible and bounded, then (S) is feasible. Suppose that (D) is feasible and bounded, and recall from Section 5.1 that any linear programme can be expressed in standard form. We may therefore apply part (i) of the current theorem with (S) replaced by (D), inferring that the dual of (D) is soluble. But by Theorem 5.3.1, the dual of (D) is (S). Thus (S) is soluble and therefore feasible. QED

An easy way to remember the Duality Theorem is to consider the following table.

		(D)		
		Feasible Bounded	Feasible Unbounded	Infeasible
(S)	Feasible Bounded	*		
	Feasible Unbounded			*
	Infeasible		*	*

We know that (S) is either feasible and bounded, feasible and unbounded, or infeasible, and that a similar taxonomy applies to (D). Thus there might seem to be nine possible cases for (S) and (D) taken together. But the Duality Theorem tells us that only four of these nine cases can occur, namely those marked * in the table. Theorem 1 also tells us that if (S) and (D) are both feasible, then both are soluble, with the same solution value.

There is one slight loose end here, concerning the South-East corner of the table above. We have not yet demonstrated that (S) and (D) can both be infeasible. The following example settles this.

Example Let $m = n = 2$ and let

$$\mathbf{A} = \begin{pmatrix} 1 & 0 \\ 0 & -1 \end{pmatrix}, \quad \mathbf{b} = \mathbf{c} = \begin{pmatrix} -1 \\ 1 \end{pmatrix}.$$

Then the constraints of (S) include

$$x_1 \leq -1 \quad \text{and} \quad x_1 \geq 0$$

so (S) is infeasible. The constraints of (D) include

$$-p_2 \geq 1 \quad \text{and} \quad p_2 \geq 0$$

so (D) is also infeasible.

The Duality Theorem has two very important corollaries. The first is the Existence Theorem of Linear Programming, mentioned in Section 5.2. The second concerns the optimality condition of Section 5.3.

THEOREM 2 (Existence Theorem)
Any feasible and bounded linear programme is soluble.
PROOF By part (i) of Theorem 1, (S) is soluble if it is feasible and bounded. The general result now follows from the fact that any linear programme may be expressed in standard form. QED

THEOREM 3 (Complementarity Theorem)
Let $\mathbf{x}$ and $\mathbf{p}$ be feasible for (S) and (D) respectively. Then the following three statements are equivalent.
(α) $\mathbf{x}$ and $\mathbf{p}$ are solution vectors for (S) and (D) respectively.
(β) $\mathbf{c}^{\mathrm{T}}\mathbf{x} = \mathbf{b}^{\mathrm{T}}\mathbf{p}$.
(γ) $(\mathbf{x} : \mathbf{b} - \mathbf{A}\mathbf{x})//(\mathbf{A}^{\mathrm{T}}\mathbf{p} - \mathbf{c} : \mathbf{p})$.
PROOF Given feasibility of $\mathbf{x}$ and $\mathbf{p}$, the equivalence of (β) and (γ) is Theorem 5.3.4 and the fact that (β) implies (α) is Theorem 5.3.3. We also know, from the Duality Theorem, that if (S) and (D) are soluble they have the same solution value. Thus (α) implies (β). QED

Exercise

Let (S) be a standard-form linear programme and (D) its dual. Which of the following propositions are always true?
(i) If (D) is feasible then (S) is feasible.
(ii) If (S) and (D) are both feasible then (D) is soluble.
(iii) If (D) is feasible and unbounded then (S) is infeasible.

7.2 Feasible bases

In this section we introduce the concept of a feasible basis, which plays an important role in more traditional treatments of linear programming; we describe the revised simplex method, which is the basis of most machine routines for linear programming; and we prove a theorem on uniqueness of solutions to linear programmes. This is all very interesting, but the only part we shall need in subsequent sections is the Uniqueness Theorem and the two examples which illustrate it. So we start with these. The proof of the Uniqueness Theorem comes at the end of the section.

UNIQUENESS THEOREM If there exists a scheme for (S) and (D) of the type (i) below, there is exactly one solution vector for (S). If there exists a scheme of type (ii) there is exactly one solution vector for (D).

$$\begin{array}{c|cccc}
 & + & + & \cdots & + \\
\hline
\oplus & & & & \\
\oplus & & & & \\
\vdots & & & & \\
\oplus & & & &
\end{array}
\qquad\qquad
\begin{array}{c|cccc}
 & \oplus & \oplus & \cdots & \oplus \\
\hline
+ & & & & \\
+ & & & & \\
\vdots & & & & \\
+ & & & &
\end{array}$$

(i) (ii)

Example 1 Consider programmes (S) and (D) where

$$\mathbf{A} = \begin{pmatrix} 2 & 4 & 1 & 1 \\ 1 & 0 & 1 & 5 \\ 3 & 1 & 3 & 0 \end{pmatrix}, \quad \mathbf{b} = \begin{pmatrix} 40 \\ 30 \\ 75 \end{pmatrix}, \quad \mathbf{c} = \begin{pmatrix} 4 \\ 3 \\ 3 \\ 2 \end{pmatrix}.$$

Applying the simplex method, we end with the scheme

$$\begin{array}{c|c|cccc}
 & & 5 & 2 & 7 & 6 \\
\hline
 & 91 & 1 & \frac{8}{5} & \frac{3}{5} & \frac{1}{5} \\
\hline
1 & 14 & 1 & \frac{56}{15} & -\frac{4}{15} & -\frac{1}{5} \\
4 & 1 & 0 & -\frac{1}{15} & -\frac{1}{15} & \frac{1}{5} \\
3 & 11 & -1 & -\frac{17}{5} & \frac{3}{5} & \frac{1}{5}
\end{array}$$

This scheme satisfies both (i) and (ii). Hence $(14\ 0\ 11\ 1)^T$ is the unique solution vector for (S) and $(1\ \frac{1}{5}\ \frac{3}{5})^T$ is the unique solution vector for (D).

Example 2 Let the data be as in Example 1, save that the second component of **b** is now 25 rather than 30. Pivoting as before we obtain a final scheme similar to that of Example 1, save that the left border is now

	90	
1	15	
4	0	
3	10	

This scheme satisfies (i) but not (ii): $(15\ \ 0\ \ 10\ \ 0)^{\mathrm{T}}$ is the unique solution vector for (S) and $\mathbf{p}^1 = (1\ \ \frac{1}{5}\ \ \frac{3}{5})^{\mathrm{T}}$ is a solution vector (but not necessarily the only one) for (D).

Indeed, by pivoting on the entry $-\frac{1}{15}$ in the row labelled '4' and the column labelled '7' we obtain a new 'Case 1' scheme:

		5	2	4	6
	90	1	1	9	2
1	15	1	4	−4	−1
7	0	0	1	−15	−3
3	10	−1	−4	9	2

Thus another solution vector for (D) is $\mathbf{p}^2 = (1\ \ 2\ \ 0)^{\mathrm{T}}$. It is easy to check that no other pivots are possible. The set of all solution vectors for (D) consists of $\mathbf{p}^1$, $\mathbf{p}^2$ and (by convexity) all vectors of the form $\alpha\mathbf{p}^1 + (1 - \alpha)\mathbf{p}^2$, where $0 < \alpha < 1$.

We now turn to the main topics of this section. Let Σ be a row feasible scheme for (S) and (D), with entries (α_{ij}) and labels (π_k). Consider the following variant of the simplex method. If we are in Case 1, we go home. If not, we choose some negative α_{0s} (generally the largest in absolute value) and look at α_{is} for $i = 1, \ldots, m$. If none of these is positive, we are in Case 2 and go home. If some α_{is} is positive we choose a pivot entry by the usual comparison of ratios and pivot-and-switch.

Notice that this is a *variant* of the simplex method, since it does not scan all the columns of Σ to check whether we are in Case 2. We call it Simplex B, thereby indicating that it is most useful when we are pretty sure that (S) is Bounded.

It turns out that Simplex B can be given a computational layout that is very different from ordinary pivot-and-switch, and often much more efficient (at any rate for a machine). To explain this we must first attend to the rather tedious task of expressing the entries of Σ in terms of *matrix* operations on the original data $\mathbf{A}$, $\mathbf{b}$, $\mathbf{c}$.

The equation system implicit in (S) may be written

$$\begin{pmatrix} \mathbf{c}^{\mathrm{T}} & \mathbf{0} \\ \mathbf{A} & \mathbf{I}_m \end{pmatrix}\begin{pmatrix} \mathbf{x} \\ \mathbf{y} \end{pmatrix} = \begin{pmatrix} \theta \\ \mathbf{b} \end{pmatrix} \tag{1}$$

where θ is to be maximised subject to $(\mathbf{x} : \mathbf{y}) \geq \mathbf{0}$. Let us reorder the columns of the coefficient matrix in (1) in the order $\pi_1, \ldots, \pi_{n+m}$ (the labels of Σ) and adjust the vector of variables accordingly. Thus (1) is rewritten as

$$\begin{pmatrix} \mathbf{u}^{\mathrm{T}} & \mathbf{v}^{\mathrm{T}} \\ \mathbf{G} & \mathbf{F} \end{pmatrix}\begin{pmatrix} \tilde{\mathbf{x}} \\ \tilde{\mathbf{y}} \end{pmatrix} = \begin{pmatrix} \theta \\ \mathbf{b} \end{pmatrix} \tag{2}$$

where u_1 denotes component π_1 of the $(n+m)$-vector $(\mathbf{c} : \mathbf{0})$, f_{23} denotes the $(2, \pi_{n+3})$ entry of $(\mathbf{A}, \mathbf{I}_m)$, $\tilde{y}_4$ denotes component π_{n+4} of $(\mathbf{x} : \mathbf{y})$, and so on.

Ignore for the moment the top equation of (2) and concentrate on the remaining system. We know from Section 3.3 that a solution to this system is

$$\tilde{\mathbf{x}} = \mathbf{0}, \quad \tilde{y}_i = \alpha_{i0} \quad \text{for all } i = 1, \ldots, m.$$

Indeed, this is the *only* solution for which $\tilde{\mathbf{x}} = \mathbf{0}$. Thus the $m \times m$ matrix $\mathbf{F}$ in (2) is nonsingular and is therefore a basis for R^m. By row-feasibility, $\mathbf{b}$ is a linear combination of the columns of $\mathbf{F}$ with non-negative weights. For this reason, $\mathbf{F}$ is called a *feasible basis* with respect to the problem (S).

Now Σ describes a row system equivalent to (2) in which $\tilde{\mathbf{y}}$ does not appear in the top equation, and appears with coefficient $\mathbf{I}_m$ in the remaining system. Operating on (2) using the nonsingularity of $\mathbf{F}$, we may write this equivalent row system as

$$\begin{aligned} &\theta = \mathbf{u}^{\mathrm{T}}\mathbf{F}^{-1}\mathbf{b} - (\mathbf{u}^{\mathrm{T}}\mathbf{F}^{-1}\mathbf{G} - \mathbf{v}^{\mathrm{T}})\tilde{\mathbf{x}}, \\ &\mathbf{F}^{-1}\mathbf{b} = \tilde{\mathbf{y}} + \mathbf{F}^{-1}\mathbf{G}\tilde{\mathbf{x}}. \end{aligned}$$

Recalling the sign convention for the top rows of schemes, we may write the entries of Σ as the matrix

$$\begin{pmatrix} \mathbf{u}^{\mathrm{T}}\mathbf{F}^{-1}\mathbf{b} & \mathbf{u}^{\mathrm{T}}\mathbf{F}^{-1}\mathbf{G} - \mathbf{v}^{\mathrm{T}} \\ \mathbf{F}^{-1}\mathbf{b} & \mathbf{F}^{-1}\mathbf{G} \end{pmatrix}.$$

What information on this matrix do we need in order to apply Simplex B? Clearly we need the left column and the top row. Also, if we want to consider the column of Σ labelled π_s as a candidate for pivoting, we need column s of $\mathbf{F}^{-1}\mathbf{G}$. Given this information, we find a pivot entry and pivot-and-switch. This gives us a new scheme and hence a new feasible basis.

So far, we have merely repeated what we already knew and derived a rather complicated expression for Σ. But this expression suggests an interesting alternative. Suppose that we do not know Σ but do know its labels, the vector $\mathbf{F}^{-1}\mathbf{b}$ *and* the inverted feasible basis $\mathbf{F}^{-1}$. Then the other things we need for a step of Simplex B can be calculated using this information and the original data $\mathbf{A}$ and $\mathbf{c}$. Specifically, $\mathbf{u}^{\mathrm{T}}\mathbf{F}^{-1}\mathbf{G}$ can be computed as $(\mathbf{u}^{\mathrm{T}}\mathbf{F}^{-1})\mathbf{G}$; and if column s of $\mathbf{F}^{-1}\mathbf{G}$ is to be singled out it can be computed as

$$\mathbf{d} = \mathbf{F}^{-1}(\text{column } s \text{ of } \mathbf{G}).$$

Neither of these calculations requires the full matrix $\mathbf{F}^{-1}\mathbf{G}$.

The pivot step (given s) can be accomplished as follows. A pivot row r is chosen as usual, by comparing the positive components of $\mathbf{d}$ with the corresponding components of $\mathbf{F}^{-1}\mathbf{b}$. The new $\mathbf{F}^{-1}$ and $\mathbf{F}^{-1}\mathbf{b}$ are obtained from the old ones by doing one Gauss–Jordan (DAWG) pivot on the $(r, m+2)$ entry of

$$(\mathbf{F}^{-1} \quad \mathbf{F}^{-1}\mathbf{b} \quad \mathbf{d})$$

and then dropping the last column. The new labels are obtained from the old ones by the usual switch. We are now ready for the next iteration.

The procedure is known as the *revised simplex method*. The main difference between it and the simplex method lies in the size of the matrices we manipulate. The hard work in the simplex method lies in successive updating, via Tucker pivoting, of the $m \times n$ matrix $\mathbf{F}^{-1}\mathbf{G}$. The hard work in the revised simplex method lies in successive updating, via Gauss–Jordan pivoting, of the $m \times m$ matrix $\mathbf{F}^{-1}$. So the revised method has an obvious advantage in cases where n is much greater than m: dual methods are available for the opposite case. The other point in favour of the revised method is that it builds up its successive $\mathbf{F}^{-1}$ matrices starting with $\mathbf{I}_m$: the initial $\mathbf{F}^{-1}\mathbf{G}$ for the simplex method is $\mathbf{A}$, which can be quite nasty.

The revised simplex method is trickier to implement by hand than the simplex method, but the advantages just discussed make it the basis of most computer routines for linear programming. A modern variant of the revised simplex method does not even calculate $\mathbf{F}^{-1}$ explicitly. Recall that what we are really interested in is not $\mathbf{F}^{-1}$ itself but the vectors

$$\mathbf{F}^{-1}\mathbf{b}, \quad (\mathbf{F}^{\mathrm{T}})^{-1}\mathbf{u} \quad \text{and} \quad \mathbf{d} = \mathbf{F}^{-1}\mathbf{g}_s$$

where $\mathbf{g}_s$ is column s of $\mathbf{G}$. Given $\mathbf{F}$, these can be computed by elimination and substitution. Of course, starting Gaussian elimination afresh each time we choose a new $\mathbf{F}$ is much harder work than doing one DAWG pivot step, but updating methods which avoid this problem have been developed. For a clear description of such a procedure see Section 3.9 of Luenberger (1973).

We now return to the theory and use the partitioned-matrix expression for Σ to prove the Uniqueness Theorem.

Let $(\mathbf{x} : \mathbf{y})$ and $(\mathbf{q} : \mathbf{p})$ be $(n + m)$-vectors such that

$$\mathbf{A}\mathbf{x} + \mathbf{y} = \mathbf{b} \quad \text{and} \quad \mathbf{A}^{\mathrm{T}}\mathbf{p} - \mathbf{q} = \mathbf{c}. \tag{3}$$

By the Complementarity Theorem, $\mathbf{x}$ and $\mathbf{p}$ are solution vectors for (S) and (D) respectively if and only if $(\mathbf{x} : \mathbf{y})$ and $(\mathbf{q} : \mathbf{p})$ are non-negative and complementary.

Now let $(\tilde{\mathbf{x}} : \tilde{\mathbf{y}})$ and $(\tilde{\mathbf{q}} : \tilde{\mathbf{p}})$ denote the vectors $(\mathbf{x} : \mathbf{y})$ and $(\mathbf{q} : \mathbf{p})$ with their components rearranged in the order $\pi_1, \ldots, \pi_{n+m}$. Then (3) is equivalent to the equations

$$\tilde{\mathbf{A}}\tilde{\mathbf{x}} + \tilde{\mathbf{y}} = \mathbf{b} \quad \text{and} \quad \tilde{\mathbf{A}}^{\mathrm{T}}\tilde{\mathbf{p}} - \tilde{\mathbf{q}} = \tilde{\mathbf{c}},$$

where $\tilde{\mathbf{A}} = \mathbf{F}^{-1}\mathbf{G}, \quad \tilde{\mathbf{b}} = \mathbf{F}^{-1}\mathbf{b} \quad \text{and} \quad \tilde{\mathbf{c}} = \mathbf{v} - \tilde{\mathbf{A}}^{\mathrm{T}}\mathbf{u}$.

To say that Σ is of the type designated (i) in the Uniqueness Theorem is to say that all components of $\tilde{\mathbf{b}}$ are non-negative and all components of $\tilde{\mathbf{c}}$ are *negative*. This implies that *a* solution for (D) is given by $\tilde{\mathbf{p}} = \mathbf{0}, \tilde{\mathbf{q}} = -\tilde{\mathbf{c}}$. But since this $\tilde{\mathbf{q}}$ has only positive components, complementarity requires that *every* $(\tilde{\mathbf{x}} : \tilde{\mathbf{y}})$ which is optimal for (S) must satisfy $\tilde{\mathbf{x}} = \mathbf{0}$, whence $\tilde{\mathbf{y}} = \tilde{\mathbf{b}}$. This uniquely determines the solution vector for (S).

To say that Σ is of type (ii) is to say that all components of $\tilde{\mathbf{b}}$ are positive and all components of $\tilde{\mathbf{c}}$ are non-positive. This implies that *a* solution for (S) is given by $\tilde{\mathbf{x}} = \mathbf{0}$, $\tilde{\mathbf{y}} = \tilde{\mathbf{b}}$. But since this $\tilde{\mathbf{y}}$ has only positive components, complementarity requires that *every* $(\tilde{\mathbf{q}} : \tilde{\mathbf{p}})$ which is optimal for (D) must satisfy $\tilde{\mathbf{p}} = \mathbf{0}$, whence $\tilde{\mathbf{q}} = -\tilde{\mathbf{c}}$. This uniquely determines the solution vector for (D).

7.3 Marginal valuation

The branch of linear programming known as *sensitivity analysis* is concerned with the question: what happens to the solution value of (S) and (D) when some or all of the entries of **A**, **b** and **c** are slightly altered? We shall consider only variations in **b**: as we shall show, this leads to a result of economic interest.

Suppose that (S) and (D) are soluble. To get some handle on what happens when **b** varies, it is helpful to consider variations in a particular direction. So fix some m-vector **d** and consider, for each positive ε, the following pair of linear programmes:

(S_ε) maximise $\mathbf{c}^T\mathbf{x}$ subject to $\mathbf{A}\mathbf{x} \leq \mathbf{b} + \varepsilon\mathbf{d}$ and $\mathbf{x} \geq \mathbf{0}$;
(D_ε) minimise $(\mathbf{b} + \varepsilon\mathbf{d})^T\mathbf{p}$ subject to $\mathbf{A}^T\mathbf{p} \geq \mathbf{c}$ and $\mathbf{p} \geq \mathbf{0}$.

Notice that (D_ε) is the dual of (S_ε) and has the same constraints as (D). Since (D) is assumed soluble, (D_ε) is feasible. Thus for any given positive ε there are just two possibilities:

either (S_ε) and (D_ε) are both soluble
or (S_ε) is infeasible and (D_ε) is feasible and unbounded.

In the first case, we denote the common solution value of (S_ε) and (D_ε) by $f(\varepsilon)$. By analogy, we denote the solution value of (S) by $f(0)$.

It turns out that the key to the behaviour of (S_ε) and (D_ε) lies in the following linear programme.

(M) minimise $\mathbf{d}^T\mathbf{p}$ subject to **p** being a solution vector for (D).

We show first that this *is* a linear programme. An m-vector **p** is a solution vector for (D) if and only if it is feasible and $\mathbf{b}^T\mathbf{p} \leq f(0)$, in which case $\mathbf{b}^T\mathbf{p} = f(0)$. Thus (M) is the linear programme

$$\text{minimise } \mathbf{d}^T\mathbf{p} \qquad \text{subject to } \mathbf{A}^T\mathbf{p} \geq \mathbf{c},\ -\mathbf{b}^T\mathbf{p} \geq -f(0) \text{ and } \mathbf{p} \geq \mathbf{0}.$$

Since (D) is soluble, (M) is feasible.

We now state the theorem which connects (M) with (S_ε). The proof is in Appendix B.

MARGINAL VALUE THEOREM If (M) is unbounded then (S_ε) is infeasible for all positive ε.

If (M) is soluble with solution value λ then

(i) (S_ε) is soluble for some positive ε,

(ii) $f(\varepsilon) \leq f(0) + \varepsilon\lambda$ for all such ε,
(iii) $f(\varepsilon) = f(0) + \varepsilon\lambda$ for all sufficiently small positive ε.

To see how this theorem is applied, and to explain its name, we discuss an economic example.

Consider a firm which has on hand m fixed factors of production (indexed $i = 1, \ldots, m$) in positive quantities $b_1, \ldots, b_m$. We assume that all the b_i are positive and that these quantities have been paid for. Using these factors, and possibly also variable inputs which the firm can hire or fire according to its needs, the firm can produce any or all of n kinds of output (indexed $j = 1, \ldots, n$). We make the following assumptions.

(I) There exists an $m \times n$ matrix

$$\mathbf{A} = \begin{pmatrix} a_{11} & \cdots & a_{1n} \\ & \vdots & \\ a_{m1} & \cdots & a_{mn} \end{pmatrix}$$

such that for any integer $j = 1, \ldots, n$ and any non-negative number ξ, production of ξ units of output j requires ξa_{ij} units of factor i for each $i = 1, \ldots, m$.

(II) There exists an n-vector

$$\mathbf{c} = (c_1 \cdots c_n)^{\mathrm{T}}$$

such that for any integer $j = 1, \ldots, n$ and any non-negative number ξ, production of ξ units of output j yields a profit of ξc_j units of money.

(III) The firm can leave idle as much as it wishes of any factor, without cost.

Assumption (I) states that there are *constant returns to scale*: factor requirements per unit of output are independent of the scale of output. It also states that the factor requirements for producing one unit of each output are given numbers. This assumption is known as *fixed coefficients of production*, and may appear to contradict the basic economic principle that there is more than one way to skin a cat. Notice, however, that this presumed absence of substitution possibilities is more a matter of interpretation than of substance: the mathematics applies equally well when j indexes different techniques for producing the same kind of output. For purposes of exposition we shall stay with our original interpretation, with j indexing physically different outputs.

Assumption (II) is known as *price-taking behaviour*: we assume that the firm is sufficiently small relative to the market that its scale of operations does not affect the prices of its outputs, or of any variable inputs required to produce them. Notice that 'profits' in (II) are 'short-period' profits, namely revenue *less* cost of variable inputs; the costs of procuring the quantities $b_1, \ldots, b_m$ of the fixed factors are sunk costs and do not affect the calculation of profit.

Suppose now that the firm produces x_j units of output j, for $j = 1, \ldots, n$. Then total profit is

$$c_1 x_1 + c_2 x_2 + \cdots + c_n x_n.$$

For $i = 1, \ldots, m$ the quantity of factor i which is actually used is

$$a_{i1} x_1 + a_{i2} x_2 + \cdots + a_{in} x_n.$$

We now appeal to (III), which is known as the assumption of *free disposal*. This says that for each i, the total quantity of factor i used may be equal to *or less than* the available quantity b_i. Thus the constraints facing the firm are

$$a_{i1} x_1 + \cdots + a_{in} x_n \leq b_i \quad (i = 1, \ldots, m),$$
and $x_j \geq 0 \quad (j = 1, \ldots, n).$

Suppose that the firm wishes to maximise profit subject to these constraints, by suitable choice of the output levels $x_1, \ldots, x_n$. It is then faced with the programme (S). Assuming that the firm's problem is soluble, we have a solution value, say v_0. This is the maximal profit attainable, given the availability of factors.

Now consider what happens to profit when the quantity available of one factor changes. Let θ be a real number and let the quantity of a particular factor i change from b_i to $b_i + \theta$, *all the other factor quantities remaining unchanged*. Provided $\theta > -b_i$, the firm's programme remains feasible. Let the new maximal profit be $v(i, \theta)$.

We can now apply the Marginal Value Theorem to evaluate $v(i, \theta)$ for suitably small $|\theta|$. Let λ_i and μ_i denote respectively the *minimal* and *maximal* values of p_i for all solution vectors $\mathbf{p}$ for the dual programme (D). Let $\mathbf{u}_i$ be the ith column of the identity matrix of order m. By part (iii) of the theorem with $\mathbf{d} = \mathbf{u}_i$,

$$v(i, \varepsilon) = v_0 + \varepsilon\lambda_i \quad \text{for all sufficiently small positive } \varepsilon.$$

By the same result, with $\mathbf{d} = -\mathbf{u}_i$,

$$v(i, -\varepsilon) = v_0 - \varepsilon\mu_i \quad \text{for all sufficient small positive } \varepsilon.$$

Summarising, we see that for all sufficiently small positive ε,

$$(v(i, \varepsilon) - v_0)/\varepsilon = \lambda_i \quad \text{and} \quad (v_0 - v(i, -\varepsilon))/\varepsilon = \mu_i .$$

Thus λ_i is the *marginal profitability* of factor i for *upward variations*, and μ_i is the marginal profitability of factor i for *downward variations*. A convenient symbolism for this is

$$\lambda_i = (\partial v/\partial b_i)^+, \quad \mu_i = (\partial v/\partial b_i)^- .$$

Since $\lambda_i \leq \mu_i$ by definition,

$$(\partial v/\partial b_i)^+ \leq (\partial v/\partial b_i)^- . \tag{1}$$

If all solution vectors for (D) have the same ith component, then $\lambda_i = \mu_i$, and we may denote their common value by $\partial v/\partial b_i$.

This result is illustrated in Figure 7.1, which graphs maximal profit as a function of b_1, keeping $b_2, \ldots, b_m$ constant at positive values.

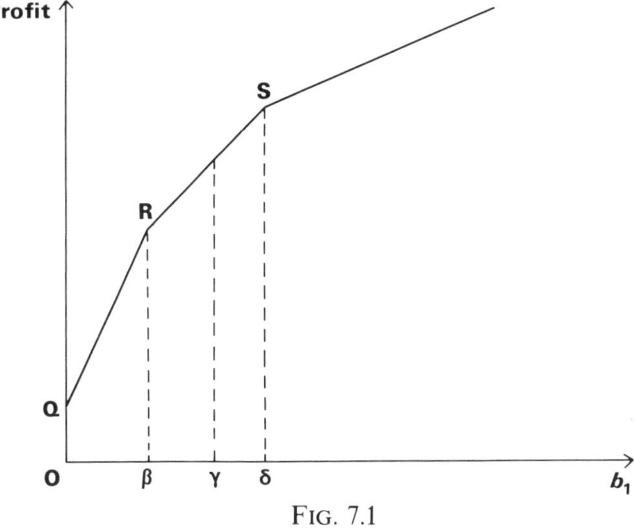

FIG. 7.1

Notice that the profit curve is piecewise linear (in other words, made up of line segments punctuated by kinks): this follows from the fact that part (iii) of the Marginal Value Theorem holds *exactly* for sufficiently small ε, not just to the first order of small quantities. There are of course cases where the point Q coincides with the origin (in other words, factor 1 is essential to profitable production): but Q cannot lie below the origin, since $\mathbf{x} = \mathbf{0}$ is feasible for (D) whenever

$\mathbf{b} \geq \mathbf{0}$. Similarly, the profit curve may not carry on increasing for ever, as in Figure 7.1, but may eventually become flat: but it will not turn down, for (by the free disposal assumption III) any vector which is feasible for (D) when $b_1 = 100$ will also be feasible when $b_1 = 200$.

We now relate the inequality (1) to the kinks in the figure. When $b_1 = \gamma$, the marginal profitability of factor 1 is the slope of the line RS. On the other hand, when $b_1 = \beta$, the marginal profitability of factor 1 for upward variations is the slope of the line RS, whereas the marginal profitability of factor 1 for downward variations is the slope of the line QR. Thus the fact that we drew Figure 7.1 with the profit curve concave (in other words, each line segment has a slope greater than the one to the right of it) was not an arbitrary choice but a consequence of the inequality (1). Similarly,

$$(\partial v/\partial b_1)^+ < (\partial v/\partial b_1)^- \quad \text{when } b_1 = \delta.$$

Readers who have studied price theory using the calculus may be interested in the relationship of the above analysis to Euler's Theorem on homogeneous functions, otherwise known as P. H. Wicksteed's rule of exhaustion of the product. Let $\mathbf{p}$ be any solution vector for (D): letting v denote the common solution value of (S) and (D), we have

$$b_1 p_1 + \cdots + b_m p_m = v. \tag{2}$$

Suppose there is only one solution vector for (D): then in this case the marginal profitability of factor i is well-defined as p_i for $i = 1, \ldots, m$. Thus (2) may be written as

$$b_1(\partial v/\partial b_1) + \cdots + b_m(\partial v/\partial b_m) = v,$$

a result highly reminiscent of Euler's Theorem.

Now consider the case where there is more than one solution vector for (D). Let $\mathbf{p}^1$ and $\mathbf{p}^2$ be two different solution vectors for (D). Then $\mathbf{b}^{\mathrm{T}}\mathbf{p}^1 = \mathbf{b}^{\mathrm{T}}\mathbf{p}^2 = v$. Since all components of $\mathbf{b}$ are positive and $\mathbf{p}^1 \neq \mathbf{p}^2$, we *cannot* have $\mathbf{p}^1 \geq \mathbf{p}^2$. This result has the following implications for the upward marginal profitabilities $\lambda_1, \ldots, \lambda_m$: whereas, for any given i, λ_i is the ith component of *some* solution vector for (D), the numbers $\lambda_1, \ldots, \lambda_m$ cannot all be the components of the *same* solution vector for (D). Thus in this case

$$b_1(\partial v/\partial b_1)^+ + \cdots + b_m(\partial v/\partial b_m)^+ < v,$$
$$b_1(\partial v/\partial b_1)^- + \cdots + b_m(\partial v/\partial b_m)^- > v.$$

Example We interpret Examples 7.2.1 and 7.2.2 as profit-maximisation problems. Let

$$\mathbf{A} = \begin{pmatrix} 2 & 4 & 1 & 1 \\ 1 & 0 & 1 & 5 \\ 3 & 1 & 3 & 0 \end{pmatrix}, \quad \mathbf{b} = \begin{pmatrix} 40 \\ \beta \\ 75 \end{pmatrix}, \quad \mathbf{c} = \begin{pmatrix} 4 \\ 3 \\ 3 \\ 2 \end{pmatrix}.$$

We know that when $\beta = 30$, the solution vector for (S) is $(14\ 0\ 11\ 1)^{\mathrm{T}}$ and the only solution vector for (D) is $(1\ \frac{1}{5}\ \frac{3}{5})^{\mathrm{T}}$. Thus the firm should produce 14 units of product 1, none of product 2, 11 units of product 3 and 1 unit of product 4. The marginal profitabilities of the three factors are 1 for factor 1, $\frac{1}{5}$ for factor 2 and $\frac{3}{5}$ for factor 3.

Now consider the case where $\beta = 25$. Then the solution vector for (S) is $(15\ 0\ 10\ 0)^{\mathrm{T}}$ and the set of solution vectors for (D) consists of all 3-vectors of the form $\alpha\mathbf{p}^1 + (1 - \alpha)\mathbf{p}^2$, where $0 < \alpha < 1$ and

$$\mathbf{p}^1 = \begin{pmatrix} 1 \\ \frac{1}{5} \\ \frac{3}{5} \end{pmatrix}, \quad \mathbf{p}^2 = \begin{pmatrix} 1 \\ 2 \\ 0 \end{pmatrix}.$$

Thus, in this case

$$\begin{aligned} \partial v/\partial b_1 = 1, \quad &(\partial v/\partial b_2)^+ = \tfrac{1}{5}, \quad (\partial v/\partial b_3)^+ = 0 \\ &(\partial v/\partial b_2)^- = 2, \quad (\partial v/\partial b_3)^- = \tfrac{3}{5}. \end{aligned}$$

Let us now summarise our economic results. Consider factor 1. As above, let λ_1 be the first component of the solution vector for (D) with minimal first component. Then we know that $\lambda_1 = (\partial v/\partial b_1)^+$, the marginal profitability of factor 1 for upward variations. Suppose now that the firm can purchase more of factor 1 at a price of r_1 per unit. Then if $r_1 < \lambda_1$ it is worth buying ε more units of factor 1, for sufficiently small positive ε. It may be argued that this general result is not particularly useful, since how small 'sufficiently small' is depends on the parameters of the problem at hand. But observe that the result just stated has a converse which is rather stronger: it follows from part (ii) of the Marginal Value Theorem that if $r_1 \geq \lambda_1$ it is not worth purchasing *any* more of factor 1. So the procedure is as follows: if $r_1 \geq \lambda_1$, the firm is all right as it is; if $r_1 < \lambda_1$, the firm should recompute its problem, with factor 1 variable, to see how much more it should buy. A similar argument holds for downward variations, when the firm is considering how much of its endowment of a given factor should be *sold* at a given market price: for an illustration, see Exercise 2.

Notice that if factor 1 is already being under-utilised, it is obviously profitable to sell some of it at any positive price: in this case $\mu_1 = 0$, so that every solution vector for (D) has first component zero. This is, of course, one half of the basic complementarity result: the m-vectors $\mathbf{b} - \mathbf{Ax}$ and $\mathbf{p}$ are non-negative and complementary.

The other half states that for optimal $\mathbf{x}$ and $\mathbf{p}$, the n-vectors $\mathbf{x}$ and $\mathbf{A}^T\mathbf{p} - \mathbf{c}$ are non-negative and complementary. This has the following economic interpretation. Let $\mathbf{p}$ be any solution vector for (D) and consider this as a vector of *imputed prices* for the factors of production. We can define the imputed long-period profit per unit of output j as the jth component of $\mathbf{c} - \mathbf{A}^T\mathbf{p}$. The feasibility of $\mathbf{p}$ implies that no kind of output makes a positive profit in this sense. Complementarity implies that those kinds of output which are produced in a solution vector for (S) just break even.

Exercises

1. The function $f(\alpha, \beta)$ is defined to be the maximal value of the linear form

$$5x_1 + 6x_2 + 7x_3 + 8x_4$$

subject to the constraints

$$\begin{aligned} 5x_1 + x_2 + 3x_3 + 2x_4 &\leq \alpha \\ 3x_1 + 2x_2 + 2x_3 + 3x_4 &\leq \beta \\ 2x_1 + 6x_2 + 4x_3 + 5x_4 &\leq 10 \\ x_1 \geq 0,\ x_2 \geq 0,\ x_3 \geq 0,\ x_4 &\geq 0. \end{aligned}$$

Evaluate $f(\alpha, \beta)$
(i) when $\alpha = 3$, $\beta = 4$;
(ii) when $\alpha = 4$, $\beta = 4$.

In each case state whether the partial derivatives $\partial f/\partial\alpha$ and $\partial f/\partial\beta$ exist, and evaluate them if they do.

2. Firm A has three machines, $M1$, $M2$, $M3$. Each machine may be operated for up to 40 hours per week. Maintenance costs and depreciation on each machine are independent of the number of hours per week for which that machine is operated.

Firm A can produce a product, which sells at £100 per unit, by any of four techniques $T1$, $T2$, $T3$, $T4$. There is no cost in switching from one technique to another in the course of a week. Labour costs £3 per hour. Machine-hour and man-hour requirements, and raw-material costs per unit of product, are given for each technique by the following table:

	$T1$	$T2$	$T3$	$T4$
$M1$ (hours)	2	2	1	0
$M2$ (hours)	1	0	0	4
$M3$ (hours)	1	1	3	1
Labour (hours)	11	13	14	8
Raw materials (£)	40	41	43	44

Firm A desires to maximise profits.

(i) Find the optimal production plan.

(ii) Now suppose that Firm B offers to rent machine $M3$ from Firm A at a rental of £5 per hour for all or any part of the week. Show that Firm A should lease $M3$ to Firm B for 15 hours per week.

(iii) Suppose that Firm B's offer still stands, and that Firm C offers to rent machine $M2$ from Firm A at a rental of £6 per hour for all or any part of the week. Show that Firm A should reject this offer.

7.4 Strict Complementarity

Having given the Complementarity Theorem an economic interpretation, we discuss the mathematics of it in more detail. Throughout this section (S) and (D) are assumed soluble, but we do not assume that the components of $\mathbf{b}$ are positive or even non-negative.

The n-vectors $\mathbf{x}$ and $\mathbf{q}$ are said to be *strictly complementary* if for each $j = 1, \ldots, n$ *exactly one* of x_j and q_j is zero. If $\mathbf{x}$ and $\mathbf{p}$ are solution vectors for (S) and (D) then the $(n + m)$-vectors

$$(\mathbf{x} : \mathbf{b} - \mathbf{A}\mathbf{x}) \quad \text{and} \quad (\mathbf{A}^{\mathrm{T}}\mathbf{p} - \mathbf{c} : \mathbf{p})$$

are non-negative and complementary, but not necessarily strictly so.

To illustrate this we appeal again to Example 7.2.2. Here the unique solution vector for (S) is

$$\mathbf{x} = (15 \quad 0 \quad 10 \quad 0)^{\mathrm{T}}.$$

Two solution vectors for (D) are

$$\mathbf{p}^1 = \begin{pmatrix} 1 \\ \frac{1}{5} \\ \frac{3}{5} \end{pmatrix}, \quad \mathbf{p}^2 = \begin{pmatrix} 1 \\ 2 \\ 0 \end{pmatrix}.$$

From the schemes of the earlier example we infer that

$$\begin{aligned}\mathbf{z} &= (\mathbf{x} : \mathbf{b} - \mathbf{A}\mathbf{x}) = (15 \quad 0 \quad 10 \quad 0 \quad 0 \quad 0 \quad 0)^{\mathrm{T}},\\ \mathbf{w}^1 &= (\mathbf{A}^{\mathrm{T}}\mathbf{p}^1 - \mathbf{c} : \mathbf{p}^1) = (0 \quad \tfrac{8}{5} \quad 0 \quad 0 \quad 1 \quad \tfrac{1}{5} \quad \tfrac{3}{5})^{\mathrm{T}},\\ \mathbf{w}^2 &= (\mathbf{A}^{\mathrm{T}}\mathbf{p}^2 - \mathbf{c} : \mathbf{p}^2) = (0 \quad 1 \quad 0 \quad 9 \quad 1 \quad 2 \quad 0)^{\mathrm{T}}.\end{aligned}$$

Obviously $\mathbf{z}$ and $\mathbf{w}^1$ are non-negative and complementary; but notice that the fourth component of each of these 7-vectors is zero. Similarly, $\mathbf{z}$ and $\mathbf{w}^2$ are non-negative and complementary and the seventh component of each is zero.

By convexity, the 3-vector

$$\mathbf{p}^3 = \tfrac{1}{2}\mathbf{p}^1 + \tfrac{1}{2}\mathbf{p}^2$$

is a solution vector for (D). Setting $\mathbf{w}^3 = (\mathbf{A}^{\mathrm{T}}\mathbf{p}^3 - \mathbf{c} : \mathbf{p})$ we have

$$\mathbf{w}^3 = \tfrac{1}{2}\mathbf{w}^1 + \tfrac{1}{2}\mathbf{w}^2 = (0 \quad \tfrac{13}{10} \quad 0 \quad \tfrac{9}{2} \quad 1 \quad \tfrac{11}{10} \quad \tfrac{3}{10})^{\mathrm{T}}$$

so that $\mathbf{z}$ and $\mathbf{w}^3$ are strictly complementary.

This numerical example suggests the following general conjecture: given solubility of (S) and (D) there always exists *some* pair of solution vectors $\mathbf{x}$ and $\mathbf{p}$ such that $(\mathbf{x} : \mathbf{b} - \mathbf{A}\mathbf{x})$ and $(\mathbf{A}^{\mathrm{T}}\mathbf{p} - \mathbf{c} : \mathbf{p})$ are strictly complementary. The following theorem states that this conjecture is correct.

STRICT COMPLEMENTARY THEOREM If (S) and (D) are soluble, there exist solution vectors $\bar{\mathbf{x}}$ and $\bar{\mathbf{p}}$ such that all components of the $(n + m)$-vector

$$(\bar{\mathbf{x}} + \mathbf{A}^{\mathrm{T}}\bar{\mathbf{p}} - \mathbf{c} : \mathbf{b} - \mathbf{A}\bar{\mathbf{x}} + \bar{\mathbf{p}})$$

are positive.

PROOF Let X and P denote the sets of all solution vectors for (S) and (D) respectively.

Fix an integer k such that $1 \leq k \leq m$ and let $\mathbf{d}$ denote the m-vector whose kth component is (-1) and whose other components are all zero. For each positive number ε let (S_ε) denote the linear programme

$$\text{maximise } \mathbf{c}^{\mathrm{T}}\mathbf{x} \quad \text{subject to } \mathbf{A}\mathbf{x} \leq \mathbf{b} + \varepsilon\mathbf{d} \text{ and } \mathbf{x} \geq \mathbf{0}.$$

Suppose for the moment that

$$p_k = 0 \text{ for all } \mathbf{p} \in P. \tag{1}$$

By the Marginal Value Theorem, (S_ε) has the same solution value as (S) for all sufficiently small positive ε. Let ε be small in this sense, and let $\mathbf{x}^k$ be a solution vector for (S_ε). Since $\mathbf{d} \leq \mathbf{0}$, $\mathbf{x}^k$ is also a solution vector for (S). Thus $\mathbf{x}^k$ is a member of X such that the kth component of the m-vector $(\mathbf{b} - \mathbf{A}\mathbf{x}^k)$ is at least ε. Letting $\mathbf{p}^k$ be any member of P we see that

$$\mathbf{b} - \mathbf{A}\mathbf{x}^k + \mathbf{p}^k \geq \mathbf{0}, \quad \text{with positive } k\text{th component.} \tag{2}$$

Now suppose that (1) is false. Then we may choose a member $\mathbf{p}^k$ of P with positive kth component. Letting $\mathbf{x}^k$ be any member of X, we see that (2) holds for this $\mathbf{x}^k$ and $\mathbf{p}^k$.

We have now proved that whether (1) is true or false, there exist vectors $\mathbf{x}^k \in X$, $\mathbf{p}^k \in P$ such that (2) is satisfied. But this is so for any $k = 1, \ldots, m$. Setting

$$\hat{\mathbf{x}} = m^{-1}(\mathbf{x}^1 + \cdots + \mathbf{x}^m), \quad \hat{\mathbf{p}} = m^{-1}(\mathbf{p}^1 + \cdots + \mathbf{p}^m)$$

we see that $\hat{\mathbf{x}}$ is a member of the convex set X, $\hat{\mathbf{p}}$ is a member of the convex set P, and all components of the m-vector

$$\mathbf{b} - \mathbf{A}\hat{\mathbf{x}} + \hat{\mathbf{p}}$$

are positive.

Since the dual of (D) is (S), there exist a member $\tilde{\mathbf{x}}$ of X and a member $\tilde{\mathbf{p}}$ of P such that all components of the n-vector

$$\tilde{\mathbf{x}} + \mathbf{A}^{\mathrm{T}}\tilde{\mathbf{p}} - \mathbf{c}$$

are positive. Since $\hat{\mathbf{x}}$ and $\tilde{\mathbf{x}}$ are feasible for (S), and $\hat{\mathbf{p}}$ and $\tilde{\mathbf{p}}$ are feasible for (D),

$$\mathbf{b} - \mathbf{A}\tilde{\mathbf{x}} + \tilde{\mathbf{p}} \geq 0 \quad \text{and} \quad \hat{\mathbf{x}} + \mathbf{A}^{\mathrm{T}}\hat{\mathbf{p}} - \mathbf{c} \geq \mathbf{0}.$$

Now set

$$\bar{\mathbf{x}} = (\tfrac{1}{2})(\hat{\mathbf{x}} + \tilde{\mathbf{x}}), \quad \bar{\mathbf{p}} = (\tfrac{1}{2})(\hat{\mathbf{p}} + \tilde{\mathbf{p}}).$$

By convexity, $\bar{\mathbf{x}} \in X$ and $\bar{\mathbf{p}} \in P$. Also, all components of the $(n+m)$-vector

$$\begin{pmatrix} \bar{\mathbf{x}} + \mathbf{A}^{\mathrm{T}}\bar{\mathbf{p}} - \mathbf{c} \\ \mathbf{b} - \mathbf{A}\bar{\mathbf{x}} + \bar{\mathbf{p}} \end{pmatrix}$$

are positive. QED

Exercise

Give an economic interpretation to the Strict Complementarity Theorem.

8 Topics in Linear Programming

It was shown in Section 5.1 that any linear programme can be expressed in standard form. Thus one way to solve *any* soluble linear programme is to formulate it as an (S) and use the simplex method, with an auxiliary programme if necessary to obtain an initial row feasible scheme. But we know that this may not be the quickest solution method: for example the best way to solve a diet problem is to express it as a (D) rather than as an (S). This chapter is concerned with linear programmes that are not posed explicitly in standard form, with special reference to equation constraints.

In Section 8.1 we formulate an absolutely general linear programme and state the main results on duality. These are then applied to a class of linear programmes, known as transportation problems, which have many applications in management science.

Section 8.2 concentrates on problems of the form

$$\text{maximise } \mathbf{c}^{\mathrm{T}}\mathbf{x} \quad \text{subject to } \mathbf{A}\mathbf{x} = \mathbf{b} \text{ and } \mathbf{x} \geq \mathbf{0}.$$

Some books (for example Luenberger, 1973) refer to this rather than (S) as 'standard form': hence the title of the section. Section 8.3 uses the theory of the other standard form to derive a Theorem of the Alternative for linear inequalities, known as Farkas' Lemma. This is used to prove the Finite Separation Theorem, which underlies the theory of the next chapter.

8.1 The general case

A *family* is a set whose members are themselves sets. A *partition* of a positive integer m is a family $\{M_1, \ldots, M_k\}$ of non-overlapping sets whose union is the set $\{1, \ldots, m\}$. For example, if

$$M_1 = \{1, 2, 5\}, \quad M_2 = \varnothing, \quad M_3 = \{4, 7\}, \quad M_4 = \{3, 6\},$$

then $\{M_1, M_2, M_3, M_4\}$ is a partition of 7. (We shall use this definition of a partition consistently, but the reader is warned that it is not standard; the usual definition does not allow the empty set to be a member.)

The most general linear maximisation programme that one can think of may be written as follows:

$$\begin{array}{lrl} \text{(L)} & \multicolumn{2}{l}{\text{maximise } \mathbf{c}^{\mathrm{T}}\mathbf{x} \text{ subject to}} \\ & [\mathbf{Ax}]_i \le b_i & \text{for all } i \in M_1 \\ & [\mathbf{Ax}]_i \ge b_i & \text{for all } i \in M_2 \\ & [\mathbf{Ax}]_i = b_i & \text{for all } i \in M_3 \\ & x_j \ge 0 & \text{for all } j \in N_1 \\ & x_j \le 0 & \text{for all } j \in N_2 \\ & x_j \text{ free} & \text{for all } j \in N_3 . \end{array}$$

Here $\mathbf{A}$ is an $m \times n$ matrix, $\mathbf{b}$ is an m-vector, $\mathbf{c}$ is an n-vector, $\{M_1, M_2, M_3\}$ is a partition of m, and $\{N_1, N_2, N_3\}$ is a partition of n. $[\mathbf{Ax}]_i$ denotes the ith component of the m-vector $\mathbf{Ax}$.

With (L), we associate the following linear minimisation programme:

$$\begin{array}{lrl} \text{(L*)} & \multicolumn{2}{l}{\text{minimise } \mathbf{b}^{\mathrm{T}}\mathbf{p} \text{ subject to}} \\ & [\mathbf{A}^{\mathrm{T}}\mathbf{p}]_j \ge c_j & (j \in N_1) \\ & [\mathbf{A}^{\mathrm{T}}\mathbf{p}]_j \le c_j & (j \in N_2) \\ & [\mathbf{A}^{\mathrm{T}}\mathbf{p}]_j = c_j & (j \in N_3) \\ & p_i \ge 0 & (i \in M_1) \\ & p_i \le 0 & (i \in M_2) \\ & p_i \text{ free} & (i \in M_3). \end{array}$$

When $M_2 = M_3 = N_2 = N_3 = \varnothing$, (L) reduces to (S) and (L*) to (D). Generally, we have the following theorem.

THEOREM 1 The dual of (L) is (L*) and the dual of (L*) is (L).

Like Theorem 5.3.1, this is a theorem and not a definition. It states that when (L) is expressed in standard form and the dual is taken, that dual is equivalent to (L*): the fact that the dual of (L*) is (L) then follows from Theorem 5.3.1. The details of the proof are a thoroughly tedious exercise in partitioned-matrix notation and are therefore omitted.

Theorem 1 may be summarised in the following *duality table.*

(L) object: *maximise*	(L*) object: *minimise*
'$\leq$' constraint	variable ≥ 0
'$\geq$' constraint	variable ≤ 0
'$=$' constraint	variable free
variable ≥ 0	'$\geq$' constraint
variable ≤ 0	'$\leq$' constraint
variable free	'$=$' constraint

Notice the asymmetries: for example, a non-negative variable in (L) corresponds to a '$\geq$' constraint in (L*), whereas a non-negative variable in (L*) corresponds to a '$\leq$' constraint in (L). This should not be surprising: we have built in asymmetry from the start by making (L) a maximisation programme and (L*) a minimisation programme.

In view of Theorem 1, the dual of any linear maximisation programme may be found by reading the duality table from left to right, and the dual of any linear minimisation programme may be found by reading the duality table from right to left.

Example 1 Find the dual of the linear programme

$$\begin{aligned} \text{minimise} \quad & 4p_1 + 3p_2 - 2p_3 + 4p_4 \\ \text{subject to} \quad & p_1 - 2p_2 + 7p_3 + p_4 = 1 \\ & p_1 + 8p_2 + 5p_3 - p_4 \leq 1 \\ & -p_1 + 5p_2 - p_3 + p_4 \geq 1 \\ & p_1 \geq 0,\ p_2 \text{ free},\ p_3 \geq 0,\ p_4 \leq 0. \end{aligned}$$

Reading the duality table from right to left, we see that the dual is:

$$\begin{aligned} \text{maximise} \quad & x_1 + x_2 + x_3 \\ \text{subject to} \quad & x_1 + x_2 - x_3 \leq 4 \\ & -2x_1 + 8x_2 + 5x_3 = 3 \\ & 7x_1 + 5x_2 - x_3 \leq -2 \\ & x_1 - x_2 + x_3 \geq 4 \\ & x_1 \text{ free},\ x_2 \leq 0,\ x_3 \geq 0. \end{aligned}$$

Using Theorem 1, and more tedious manipulation of partitioned matrices, one may prove the following:

The Duality Theorem (*7.1.1*) *and the Complementarity Theorem* (*7.1.3*) *remain true when* (S) *and* (D) *are replaced by* (L) *and* (L*) *respectively. From now on we refer to these generalised results as the Duality Theorem and the Complementarity Theorem respectively.*

As an application of these principles, we consider the following problem.

Let $s_1, \ldots, s_m, d_1, \ldots, d_n$ be positive numbers such that

$$s_1 + \cdots + s_m = d_1 + \cdots + d_n \tag{1}$$

Also let

$$\begin{pmatrix} c_{11} & \cdots & c_{1n} \\ & \vdots & \\ c_{m1} & \cdots & c_{mn} \end{pmatrix}$$

be a given $m \times n$ matrix and let

$$\begin{pmatrix} x_{11} & \cdots & x_{1n} \\ & \cdots & \\ x_{m1} & \cdots & x_{mn} \end{pmatrix}$$

be a matrix of mn variables. Our linear programme is:

(T) minimise $\sum_{i=1}^{m} \sum_{j=1}^{n} c_{ij}x_{ij}$ subject to

$$\sum_{j=1}^{n} x_{ij} = s_i \quad \text{for } i = 1, \ldots, m \tag{2}$$

$$\sum_{i=1}^{m} x_{ij} = d_j \quad \text{for } j = 1, \ldots, n \tag{3}$$

and $x_{ij} \geq 0$ for all i, j.

The reason for being interested in problems of this type may be illustrated by an example. Suppose that a dairy company wishes to transport t gallons of milk

from m supply points (farms) indexed $i = 1, \ldots, m$
to n demand points (dairies) indexed $j = 1, \ldots, n$.

Let s_i gallons of milk be available at i $(= 1, \ldots, m)$ and let d_j gallons of milk be required at j (= 1, ..., n). We assume that there is no overall excess supply or demand: the parameters s_i and d_j satisfy (1) with both sides equal to t.

Clearly, the company has some freedom of action as to which farm is to supply how much of which dairy's requirements. Suppose that cost of transport from farm i to dairy j is c_{ij} *per gallon* for each i and j, and that the company wishes to minimise the total cost of transportation. If, for each pair (i, j), x_{ij} gallons are transported from i to j, then the total cost is

$$\sum_{i=1}^{m} \sum_{j=1}^{n} c_{ij} x_{ij}.$$

The company wishes to minimise this expression subject to the relevant constraints on the numbers x_{ij}.

These constraints are of three types: supply constraints, demand constraints and non-negativity constraints. The supply constraints state that the total quantity of milk transported from farm i should be exactly s_i for each i: these constraints are summarised in (2). The demand constraints state that the total quantity of milk transported to dairy j should be exactly d_j for each j: these constraints are summarised in (3). The non-negativity constraints are self-explanatory. Thus the dairy company's problem is precisely (T). For this reason, we refer to problems of the form (T) as *transportation problems*.

A matrix **X** which satisfies the constraints of (T) will be called an *allocation*. Letting t denote the value of both sides of (1), we may obtain an allocation by setting

$$x_{ij} = (s_i d_j)/t \quad \text{for all } i, j$$

Thus (T) is feasible.

Indeed, (T) is feasible *and bounded*. To see this, let **X** be any allocation. *Either* by adding up the supply constraints (2), *or* by adding up the demand constraints (3), we obtain the equation

$$\sum_{i=1}^{m} \sum_{j=1}^{n} x_{ij} = t. \tag{4}$$

Now let β be a number such that $\beta \leq c_{ij}$ for all i, j. Then by (4) and the non-negativity constraints,

$$\sum_{i=1}^{m} \sum_{j=1}^{n} c_{ij} x_{ij} \geq \beta t$$

so (T) is bounded.

We now use the duality table to write down the dual of (T). This is the following programme in $m + n$ variables

$$p_1, \ldots, p_m, \quad q_1, \ldots, q_n$$

all unrestricted in sign:

$$\text{(T*)} \quad \text{maximise} \quad \sum_{i=1}^{m} s_i p_i + \sum_{j=1}^{n} d_j q_j$$
$$\text{subject to} \quad p_i + q_j \leq c_{ij} \quad \text{for all } i, j.$$

Since (T) is feasible and bounded, we infer from the Duality Theorem that (T) and (T*) are soluble, with the same solution value. Applying the Complementarity Theorem to (T) and (T*), we obtain the following theorem.

THEOREM 2 Let $\mathbf{X}$ be an $m \times n$ matrix and $(\mathbf{p} : \mathbf{q})$ an $(m + n)$-vector. Then the following two statements are equivalent.
(α) $\mathbf{X}$ and $(\mathbf{p} : \mathbf{q})$ are solutions for (T) and (T*) respectively.
(β) $\mathbf{X}$ is feasible for (T) and

$$p_i + q_j \leq c_{ij} \quad \text{for all } i, j$$

with equality if $x_{ij} > 0$.

Theorem 2 is one of the two main building blocks in the standard algorithm for solving transportation problems. We shall have a little more to say about this at the end of the next section.

8.2 The other standard form

Let $\mathbf{A}$ be an $m \times n$ matrix, $\mathbf{b}$ an m-vector and $\mathbf{c}$ an n-vector. We consider the linear programme

$$\text{(SE)} \quad \text{maximise } \mathbf{c}^{\mathrm{T}}\mathbf{x} \quad \text{subject to } \mathbf{A}\mathbf{x} = \mathbf{b} \text{ and } \mathbf{x} \geq \mathbf{0}.$$

From the duality table of the last section we can read off the dual to (SE). This is the linear programme

$$\text{(DF)} \quad \text{minimise } \mathbf{b}^{\mathrm{T}}\mathbf{p} \quad \text{subject to } \mathbf{A}^{\mathrm{T}}\mathbf{p} \geq \mathbf{c}.$$

In (DF) the components of $\mathbf{p}$ are *free*.

Notice that (SE) can be interpreted as a profit maximisation problem without free disposal. This gives economic content to the statement that the components of $\mathbf{p}$ are unrestricted in sign in the dual

(DF): if the show is scored for string quartet and the musicians' union insists on the presence of a drummer with full kit, there is no reason why the drummer or his kit should have non-negative marginal profitability.

Let us find an algorithm for solving (SE) and (DF). In what follows we shall assume that $\mathbf{b} \geq \mathbf{0}$. This assumption can be made without loss of generality: if, for example, $b_1 < 0$ the first equation in (SE) can be multiplied by -1.

The method of solution is similar to the simplex method. We are interested in finding a complementary solution to

$$\mathbf{Ax} + \mathbf{y} = \mathbf{b} \quad \text{and} \quad \mathbf{A}^{\mathrm{T}}\mathbf{p} - \mathbf{q} = \mathbf{c} \tag{1}$$

with the additional property that

$$\mathbf{x} \geq \mathbf{0}, \quad \mathbf{y} = \mathbf{0} \quad \text{and } \mathbf{q} \geq \mathbf{0}.$$

Then $\mathbf{x}$ will be a solution vector for (SE) and $\mathbf{p}$ a solution vector for (DF).

Let Σ be a scheme for (1) with entries (α_{ij}) and labels (π_k). We say that Σ is (SE)-feasible if

$\alpha_{i0} \geq 0$ for $i = 1, \ldots, m$,
with $\alpha_{i0} = 0$ if the row label π_{n+i} exceeds n.

We say that Σ is (DF)-feasible if

$\alpha_{0j} \geq 0$ for all those $j = 1, \ldots, n$
whose column label π_j does not exceed n.

If we have a scheme which is both (SE)-feasible and (DF)-feasible we can read off solution vectors $\mathbf{x}$ and $\mathbf{p}$.

To find an initial (SE)-feasible scheme we use a variant of the auxiliary programme technique. Let $\mathbf{e}$ be the m-vector consisting entirely of 1's and let $\mathbf{a} = \mathbf{A}^{\mathrm{T}}\mathbf{e}$, $\beta = \mathbf{b}^{\mathrm{T}}\mathbf{e}$. Thus $\mathbf{a}$ is the n-vector whose jth component, for $j = 1, \ldots, n$, is the sum of the entries in the jth column of $\mathbf{A}$, and

$$\beta = b_1 + \cdots + b_m.$$

We take as our auxiliary programme the problem

$$\text{maximise } \mathbf{a}^{\mathrm{T}}\mathbf{x} - \beta \quad \text{subject to } \mathbf{Ax} \leq \mathbf{b} \text{ and } \mathbf{x} \geq \mathbf{0}.$$

This is feasible, since $\mathbf{b} \geq \mathbf{0}$ by assumption. So apart from the constant term $(-\beta)$ in the objective function, which is easily handled, the auxiliary programme is a straightforward standard-form programme which can be solved by the simplex method.

Let $\mathbf{x}$ be feasible for the auxiliary programme. By definition of $\mathbf{a}$ and β,

$$\mathbf{a}^{\mathrm{T}}\mathbf{x} - \beta = \mathbf{e}^{\mathrm{T}}(\mathbf{A}\mathbf{x} - \mathbf{b}).$$

But $\mathbf{Ax} \leq \mathbf{b}$ and all components of $\mathbf{e}$ are positive. Thus $\mathbf{a}^{\mathrm{T}}\mathbf{x} \leq \beta$, with equality if and only if $\mathbf{Ax} = \mathbf{b}$. Thus the auxiliary programme is soluble with non-positive solution value: the solution value is zero if and only if (SE) is feasible.

We now have a two-step method for solving (SE). We use the auxiliary programme as a feasibility test. If the solution value of the auxiliary programme is negative, (SE) is infeasible. If the solution value of the auxiliary programme is zero, then (SE) is feasible and we have an (SE)-feasible scheme. We then search over (SE)-feasible schemes looking for one which is also (DF)-feasible. The details of the two-step procedure are as follows.

(i) *Finding an (SE)-feasible scheme*

We start with the scheme

		1	$\cdots$	n
	0	$-c_1$	$\cdots$	$-c_n$
	β	$-a_1$	$\cdots$	$-a_n$
$n+1$				
$\vdots$	$\mathbf{b}$		$\mathbf{A}$	
$n+m$				

where β is the sum of the components of $\mathbf{b}$ and a_j is the sum of the entries in the jth column of $\mathbf{A}$.

We proceed by the simplex method, using the row immediately above the line as the objective function row. This leads us to a scheme

		π_1	$\cdots$	π_n
	α_{00}	α_{01}	$\cdots$	α_{0n}
	α_{*0}	α_{*1}	$\cdots$	α_{*n}
π_{n+1}	α_{10}	α_{11}	$\cdots$	α_{1n}
$\vdots$	$\vdots$		$\vdots$	
π_{n+m}	α_{m0}	α_{m1}	$\cdots$	α_{mn}

such that $\alpha_{i0} \geq 0$ for $i = 1, \ldots, m$,
$\alpha_{*j} \geq 0$ for $j = 1, \ldots, n$,
$\alpha_{*0} \leq 0$.

If $\alpha_{*0} < 0$ then (SE) is infeasible.

If $\alpha_{*0} = 0$ then the auxiliary programme has solution value zero. In this case $\alpha_{i0} = 0$ for all i such that $\pi_{n+i} > n$; so by striking out the row immediately above the line, we obtain an (SE)-feasible scheme.

(ii) *Working with (SE)-feasible schemes*

Suppose we have an (SE)-feasible scheme. Then there are three cases to consider.

Case 1 Each column with label not exceeding n has non-negative top entry.

In this case the scheme is (DF)-feasible so we may read off solution vectors for (SE) and (DF).

Case 2 There exists a column, with label not exceeding n, whose top entry is negative and whose other entries are all non-positive, with zero entries for rows whose labels exceed n.

In this case (SE) is feasible and unbounded and (DF) is infeasible.

Case 3 Neither Case 1 nor Case 2 occurs.

In this case we may choose a pivot column s such that $\pi_s \leq n$ and $\alpha_{0s} < 0$. If for some i we have $\pi_{n+i} > n$ and $\alpha_{i,s} \neq 0$, we choose any such i as the pivot row. Otherwise we choose the pivot row as in the simplex method.

The procedure is repeated until Case 1 or Case 2 is reached.

Example

Maximise x_1
subject to
$$x_1 + x_2 + x_3 = 2$$
$$x_1 + 2x_2 + 3x_3 = 4$$
$$x_1 \geq 0,\ x_2 \geq 0,\ x_3 \geq 0.$$

The auxiliary programme is solved as follows:

		1	2	3
	0	−1	0	0
	−6	−2	−3	−4
4	2	1	1	1
5	4	1	2	(3)

$$
\begin{array}{c|c|ccc}
 & & 1 & 2 & 5 \\
\hline
 & 0 & -1 & 0 & 0 \\
 & -\frac{2}{3} & -\frac{2}{3} & -\frac{1}{3} & \frac{4}{3} \\
\hline
4 & \frac{2}{3} & \textcircled{\frac{2}{3}} & \frac{1}{3} & -\frac{1}{3} \\
3 & \frac{4}{3} & \frac{1}{3} & \frac{2}{3} & \frac{1}{3}
\end{array}
$$

$$
\begin{array}{c|c|ccc}
 & & 4 & 2 & 5 \\
\hline
 & 1 & \frac{3}{2} & \frac{1}{2} & -\frac{1}{2} \\
 & 0 & 1 & 0 & 1 \\
\hline
1 & 1 & \frac{3}{2} & \frac{1}{2} & -\frac{1}{2} \\
3 & 1 & -\frac{1}{2} & \frac{1}{2} & \frac{1}{2}
\end{array}
$$

We can now strike out the row immediately above the line and obtain the (SE)-feasible scheme

$$
\begin{array}{c|c|ccc}
 & & 4 & 2 & 5 \\
\hline
 & 1 & \frac{3}{2} & \frac{1}{2} & -\frac{1}{2} \\
\hline
1 & 1 & \frac{3}{2} & \frac{1}{2} & -\frac{1}{2} \\
3 & 1 & -\frac{1}{2} & \frac{1}{2} & \frac{1}{2}
\end{array}
$$

Since the column with label not exceeding 3 has a positive top entry, we are in Case 1. The solution to our programme is

$$x_1 = 1, \quad x_2 = 0, \quad x_3 = 1$$

and the solution value is 1.

The dual programme is

$$\text{minimise } 2p_1 + 4p_2 \quad \text{subject to}$$
$$p_1 + p_2 \geq 1, \quad p_1 + 2p_2 \geq 0, \quad p_1 + 3p_2 \geq 0.$$

The solution to the dual is $p_1 = \frac{3}{2}$, $p_2 = -\frac{1}{2}$ and the solution value is 1.

As a consequence of the algorithm we have the following theorem.

THEOREM 1 Let $\mathbf{A}$ be an $m \times n$ matrix, $\mathbf{b}$ an m-vector, $\mathbf{c}$ an n-vector. If the linear programme

$$\text{maximise } \mathbf{c}^{\mathrm{T}}\mathbf{x} \quad \text{subject to } \mathbf{A}\mathbf{x} = \mathbf{b} \text{ and } \mathbf{x} \geq \mathbf{0}$$

is soluble, there exists a solution vector with at most m positive components.

Luenberger (1973, p. 18) calls this result the fundamental theorem of linear programming. In our approach it is merely a by-product of the method of complementary solutions. We shall use it in Appendix D to prove the nonsubstitution theorem of input–output analysis. We apply it here to the transportation problem.

Our notation for (T) and (T*) is taken from the preceding section. Given an allocation **X** we let $K(\mathbf{X})$ denote the set of all pairs (i, j) for which $x_{ij} > 0$. Recall that each pair (i, j) symbolises the route from i to j, so $K(\mathbf{X})$ is the set of routes used by the allocation **X**. An allocation is said to be *basic* if it uses at most $m + n - 1$ routes.

THEOREM 2 For any transportation problem, there exists an optimal allocation which is basic.

PROOF Let t be the total amount transported. Recall from the last section that the equation

$$\sum_{i=1}^{m} \sum_{j=1}^{n} x_{ij} = t$$

can be obtained *either* by summing the supply constraints *or* by summing the demand constraints. Thus the $m + n$ equation constraints of (T) are not independent: we can drop any one of them without altering the nature of the problem. This means that we can formulate (T) as a linear programme in mn non-negative variables subject to only $m + n - 1$ equation constraints. The required result now follows from Theorem 1. QED

We stated in the last section that Theorem 8.1.2 is one of the two main building blocks for the transportation algorithm. Theorem 2 of this section is the other one: it says that we may restrict our search over allocations to basic ones.

The algorithm works as follows.

STEP ONE Find a basic allocation **X**.

STEP TWO Find an $(m + n)$-vector $(\mathbf{p} : \mathbf{q})$ such that

$$p_i + q_j = c_{ij} \quad \text{for all } (i, j) \text{ in } K(\mathbf{X}).$$

Since **X** is basic, we have to solve at most $m + n - 1$ equations in $m + n$ unknowns. The degree of freedom is helpful, since we can start by setting one component of **p** arbitrarily equal to zero.

If the resulting $(\mathbf{p} : \mathbf{q})$ is feasible for (T*) then, by Theorem 8.1.2, **X** is optimal for (T) and $(\mathbf{p} : \mathbf{q})$ for (T*). This motivates the next step.

STEP THREE For each route (i, j) which is not in $K(\mathbf{X})$, calculate

$$v_{ij} = c_{ij} - p_i - q_j.$$

If all such v_{ij} are non-negative then $\mathbf{X}$ is an optimal allocation, and we are finished. If some v_{ij} is negative we pass to the next step.
STEP FOUR Choose a route (h, k) for which $v_{hk} < 0$: a reasonable procedure is to choose the largest in absolute value of the negative v_{ij}. Find a new basic allocation which uses the route (h, k) and apply Steps Two and Three to it: this yields new values of p_i, q_j and v_{ij}. Continue until an allocation is reached for which all v_{ij} are non-negative.

The reader will have noticed that we have not explained how to choose the basic allocations in Steps One and Four, or how to set out the arithmetic. There are various simple ways of doing these things: for a good exposition see Chapter 10 of Gass (1975).

Exercises

1. Maximise $2x_1 + x_2 + x_3$
subject to
$$5x_1 + 6x_2 + 3x_3 = 1$$
$$8x_1 + x_2 + 6x_3 = 1$$
$$x_1 \geq 0,\ x_2 \geq 0,\ x_3 \text{ free.}$$

2. Solve the following linear programmes.
(i) Maximise $4x_1 + 3x_2 + x_3 - 7x_4$
subject to
$$6x_1 - x_2 - x_3 + 7x_4 \leq 10$$
$$4x_1 - x_2 + x_3 - 7x_4 \leq 10$$
$$x_1 + 2x_2 - x_3 + 17x_4 \leq 10$$
$$x_1 \geq 0,\ x_2 \geq 0,\ x_3 \geq 0,\ x_4 \geq 0.$$
(ii) As (i), except that the $\leq$ in the first constraint is now $=$.
(iii) As (ii), except that x_4 is now free.
(iv) The dual of (iii).

8.3 Farkas' Lemma

In Section 2.4 we proved the Theorem of the Alternative for linear equations: if $\mathbf{A}$ is an $m \times n$ matrix and $\mathbf{b}$ an m-vector then
either there exists an n-vector $\mathbf{x}$ such that $\mathbf{Ax} = \mathbf{b}$
or there exists an m-vector $\mathbf{p}$ such that $\mathbf{A}^{\mathrm{T}}\mathbf{p} = \mathbf{0}$ and $\mathbf{b}^{\mathrm{T}}\mathbf{p} \neq 0$
but not both.

Analogous theorems can be proved for systems of *linear inequalities*. Farkas' Lemma is perhaps the most commonly used theorem of this type. Two others are given in the exercises.

THEOREM 1 (Farkas' Lemma)
Let $\mathbf{A}$ be an $m \times n$ matrix, $\mathbf{b}$ an m-vector. Then
either (α) there exists an n-vector $\mathbf{x}$ such that $\mathbf{Ax} = \mathbf{b}$ and $\mathbf{x} \geq \mathbf{0}$
or (β) there exists an m-vector $\mathbf{p}$ such that $\mathbf{A}^{\mathrm{T}}\mathbf{p} \geq \mathbf{0}$ and $\mathbf{b}^{\mathrm{T}}\mathbf{p} < 0$
but not both.

PROOF Consider the programmes (SE) and (DF) of the last section in the special case where $\mathbf{c} = \mathbf{0}$. We have

$$\begin{aligned} &(\mathrm{L}) \ \text{maximise } 0 \quad \text{subject to } \mathbf{Ax} = \mathbf{b} \text{ and } \mathbf{x} \geq \mathbf{0} \\ &(\mathrm{L}^*) \ \text{minimise } \mathbf{b}^{\mathrm{T}}\mathbf{p} \quad \text{subject to } \mathbf{A}^{\mathrm{T}}\mathbf{p} \geq \mathbf{0}. \end{aligned}$$

Since its objective function is identically zero, (L) is either soluble or infeasible. Since $\mathbf{p} = \mathbf{0}$ is feasible, (L*) is either soluble or feasible-and-unbounded. So by the Duality Theorem
either (i) (L) is feasible
or (ii) (L*) is feasible and unbounded
but not both

Now (i) is identical to (α) and it remains to show that (ii) is equivalent to (β). It is obvious that (ii) implies (β). Conversely, suppose there exists some $\hat{\mathbf{p}}$ such that $\mathbf{A}^{\mathrm{T}}\hat{\mathbf{p}} \geq \mathbf{0}$ and $\mathbf{b}^{\mathrm{T}}\hat{\mathbf{p}} < 0$. By setting $\mathbf{p} = M\hat{\mathbf{p}}$ for suitably large positive M we can make $-\mathbf{b}^{\mathrm{T}}\mathbf{p}$ as large as we like while keeping $\mathbf{p}$ feasible for (L*). This implies (ii). QED

As an application of Farkas' Lemma we prove a theorem about weighted averages called the Finite Separation Theorem. This theorem will play a crucial rôle in Section 9.2. A geometrical interpretation, and the reason for the theorem's name, will be given in Section 10.1.

We say that the n-vector $\mathbf{x}_0$ is a *convex combination* (or *weighted average*) of the n-vectors $\mathbf{x}_1, \ldots, \mathbf{x}_k$ if there exist *non-negative* scalars $\alpha_1, \ldots, \alpha_k$ such that

$$\alpha_1\mathbf{x}_1 + \alpha_2\mathbf{x}_2 + \cdots + \alpha_k\mathbf{x}_k = \mathbf{x}_0 \quad \text{and} \quad \alpha_1 + \alpha_2 + \cdots + \alpha_k = 1.$$

For example,

$$\begin{pmatrix} 1 \\ -1 \end{pmatrix} = (\tfrac{1}{7})\begin{pmatrix} 5 \\ 3 \end{pmatrix} + (\tfrac{2}{7})\begin{pmatrix} 3 \\ -9 \end{pmatrix} + (\tfrac{4}{7})\begin{pmatrix} -1 \\ 2 \end{pmatrix}.$$

THEOREM 2 (Finite Separation Theorem)
If the n-vector $\mathbf{x}_0$ is *not* a convex combination of the n-vectors $\mathbf{x}_1, \ldots, \mathbf{x}_k$, there exists some $\mathbf{p}$ such that

$$\mathbf{p}^T\mathbf{x}_0 < \mathbf{p}^T\mathbf{x}_i \quad \text{for } i = 1, \ldots, k.$$

PROOF Define an $(n + 1) \times k$ matrix $\mathbf{B}$ and an $(n + 1)$-vector $\mathbf{y}$ by setting

$$\mathbf{B} = \begin{pmatrix} \mathbf{x}_1 & \mathbf{x}_2 & \cdots & \mathbf{x}_k \\ 1 & 1 & \cdots & 1 \end{pmatrix}, \quad \mathbf{y} = \begin{pmatrix} \mathbf{x}_0 \\ 1 \end{pmatrix}.$$

If $\mathbf{x}_0$ is not a weighted average of $\mathbf{x}_1, \ldots, \mathbf{x}_k$, there exists no non-negative k-vector $\mathbf{z}$ such that $\mathbf{Bz} = \mathbf{y}$. Applying Farkas' Lemma to $\mathbf{B}$ and $\mathbf{y}$, we infer the existence of an $(n + 1)$-vector $\mathbf{q}$ such that $\mathbf{B}^T\mathbf{q} \geq \mathbf{0}$ and $\mathbf{y}^T\mathbf{q} < 0$. Write $\mathbf{q} = (\mathbf{p} : \lambda)$ where $\mathbf{p}$ is an n-vector and λ a scalar. Then for $i = 1, \ldots, k$,

$$\mathbf{p}^T\mathbf{x}_i + \lambda = [\mathbf{B}^T\mathbf{q}]_i \geq 0 > \mathbf{y}^T\mathbf{q} = \mathbf{p}^T\mathbf{x}_0 + \lambda$$

so $\mathbf{p}^T\mathbf{x}_i > \mathbf{p}^T\mathbf{x}_0$. QED

Exercises

1. *Stiemke's Theorem* states that if $\mathbf{A}$ is an $m \times n$ matrix and $\mathbf{c}$ an n-vector, then
   either there exists an n-vector $\mathbf{x}$ such that $\mathbf{x} \geq \mathbf{0}$, $\mathbf{Ax} = \mathbf{0}$ and $\mathbf{c}^T\mathbf{x} > 0$
   or there exists an m-vector $\mathbf{p}$ such that $\mathbf{A}^T\mathbf{p} \geq \mathbf{c}$
   but not both.
   Prove this theorem.

2. *Motzkin's Theorem* states that if $\mathbf{A}$ is an $m \times n$ matrix and k is a positive integer not exceeding n, then
   either there exists an n-vector $\mathbf{x}$ such that $\mathbf{Ax} = \mathbf{0}$, $\mathbf{x} \geq \mathbf{0}$ and $x_j > 0$ for *some* $j \leq k$
   or there exists an m-vector $\mathbf{p}$ such that $\mathbf{A}^T\mathbf{p} \geq \mathbf{0}$ and $[\mathbf{A}^T\mathbf{p}]_j > 0$ for *all* $j \leq k$
   but not both.
   Prove this. Show also that the theorem remains true when the words 'some' and 'all' are interchanged.

3. Derive Theorem 2.4.4. from Farkas' Lemma.

9 Lagrange Multipliers

Our last three chapters are concerned with nonlinear programming, the maximisation of functions subject to constraints when the relevant functions are not necessarily linear. Unlike earlier chapters, this part of the book is concerned only with theory. The computation of maxima of nonlinear functions is a complicated business even in the unconstrained case, and we have no space to discuss it here. For an excellent discussion of computation in nonlinear programming, see Luenberger (1973).

We start this chapter with a brief discussion of differential calculus of several variables. The reader is assumed to know something about this already. In particular, he is expected to have an intuitive idea of continuity (the value taken by a continuous function responds but little to small changes in the arguments) and to know the rules of partial differentiation. He is not expected to be familiar with the representation of partial derivatives as components of vectors, and one purpose of Section 9.1 is to explain this.

Sections 9.2 and 9.3 are concerned with necessary conditions for a constrained maximum. These are expressed in terms of first derivatives and are therefore called *first-order* conditions. The original theorem of this type was established in the eighteenth century by Joseph-Louis Lagrange, who was concerned only with equation constraints. The extension to the case of weak inequality constraints was completed in the 1940s in the work of Fritz John, Harold Kuhn and A. W. Tucker. Though the classical case (all constraints equations) will be familiar to many readers, we do not start from that point;

there are certain subtleties connected with the classical case which are ignored in many elementary texts, particularly those written for economists. Instead, we start Section 9.2 with the case where all constraints are weak inequalities and work up to the case where equations are included.

9.1 Gradients

We stated earlier that this section was going to be about differential calculus. Most of it is taken up with answering the rather esoteric question, where do differentiable functions live? This is unavoidable if we are to get clear and correct statements of the main theorems.

We begin by defining some kinds of set in R^n which are associated with the Euclidean norm. Since we are now concerned with analysis, n-vectors will often be referred to as *points*, but we shall continue to represent them as columns.

Let $\mathbf{a} \in R^n$ and let δ be a positive number. The *open ball* with centre $\mathbf{a}$ and radius δ is the set

$$\{\mathbf{x} \in R^n\colon \|\mathbf{x} - \mathbf{a}\| < \delta\}$$

consisting of all points whose distance from $\mathbf{a}$ is *less* than δ. The *closed ball* with centre $\mathbf{a}$ and radius δ is the set of points whose distance from $\mathbf{a}$ is *not greater* than δ. A *ball* is an open ball or a closed ball.

The reason for 'ball' rather than 'disc' is that R^3 is significantly more complicated than R^2, so it is useful to take our terminology from the physical universe rather than the blackboard. The reason for 'ball' as opposed to 'sphere' is as follows. It is common in contemporary mathematics to define the *sphere* with centre $\mathbf{a}$ and radius δ as the set

$$\{\mathbf{x} \in R^n\colon \|\mathbf{x} - \mathbf{a}\| = \delta\}.$$

We shall use this definition, for every n.

A set in R^n is said to be *bounded* if it can be contained in a ball. For example, the set

$$\{\mathbf{x} \in R^n\colon -1 \le x_j \le 1 \quad \text{for } j = 1, \ldots, n\}$$

is bounded since it is contained in the closed ball with radius $\sqrt{n}$. The set $\{\mathbf{x} \in R^n\colon x_j \le 1 \text{ for } j = 1, \ldots, n\}$ is not bounded since it owns $(-M \; 0 \cdots 0)^{\mathrm{T}}$ for arbitrarily large positive M.

Given a set X in R^n we define its *complement* X^c to be the set of all n-vectors which are not in X. Given $\mathbf{a} \in R^n$ we say that $\mathbf{a}$ is a *boundary point* of X if every open ball with centre $\mathbf{a}$ owns at least one point of X and at least one point of X^c. The set of all boundary points of X is called the *boundary* of X and is denoted by bdry X.

Let $\mathbf{b} \in R^n$ and $\delta > 0$. Let B, $\bar{B}$ and S denote respectively the open ball, the closed ball and the sphere with radius δ and centre $\mathbf{b}$. Then

$$\text{bdry } B = S = \text{bdry } \bar{B}. \tag{1}$$

This is geometrically obvious when n is 1, 2 or 3. For $n > 3$ we cannot use geometrical intuition but we can use the properties of the norm given in Theorem 1.4.4 (properties (ii) and (iv) are the relevant ones). The next paragraph, which can be omitted without loss of continuity, sketches a proof of the left-hand equality in (1). The right-hand one is proved by a similar argument.

(For any $\mathbf{x}$ in R^n we can let $\mathbf{x}^* = \delta^{-1}(\mathbf{x} - \mathbf{b})$ and retrieve $\mathbf{x}$ from $\mathbf{x}^*$ by $\mathbf{x} = \mathbf{b} + \delta\mathbf{x}^*$. A point $\mathbf{x}$ belongs to the open ball with centre $\mathbf{a}$ and radius α if and only if $\mathbf{x}^*$ belongs to the open ball with centre $\mathbf{a}^*$ and radius α/δ. These remarks imply that there is no loss of generality in assuming that $\mathbf{b} = \mathbf{0}$ and $\delta = 1$. So let B be the set of all $\mathbf{x}$ such that $\|\mathbf{x}\| < 1$ and let S be the set of all $\mathbf{x}$ such that $\|\mathbf{x}\| = 1$. If $\|\mathbf{x}\| > 1$, the open ball with centre $\mathbf{x}$ and radius $\|\mathbf{x}\| - 1$ owns no point of B. if $\|\mathbf{x}\| < 1$ the open ball with centre $\mathbf{x}$ and radius $1 - \|\mathbf{x}\|$ owns no point of B^c. Thus if $\mathbf{x}$ is not in S, $\mathbf{x}$ is not a boundary point of B. If $\|\mathbf{x}\| = 1$ then any open ball with centre $\mathbf{x}$ owns $(1 - \lambda)\mathbf{x}$ and $(1 + \lambda)\mathbf{x}$ for suitably small positive λ, so $\mathbf{x}$ is a boundary point of B.)

We say that a set is *open* if it owns none of its boundary points, *closed* if it owns all of them. It follows from (1) that open balls are open sets and closed balls are closed sets. Since $(X^c)^c = X$ for any set X, the boundary of a set is the boundary of its complement: it follows that a set is closed if and only if its complement is open.

The *non-negative orthant* of R^n is the set of all non-negative n-vectors, and is denoted by R^n_+. The boundary of R^n_+ is the set

$$\{\mathbf{x} \in R^n_+ : x_j = 0 \text{ for some } j\}.$$

We define an n-vector $\mathbf{x}$ to be *positive* ($\mathbf{x} > \mathbf{0}$) if all its components are positive: the set of all such vectors is the *positive orthant* P^n. The boundary of P^n is the same as the boundary of R^n_+: notice that all members of this common boundary are non-negative and none is positive, so R^n_+ is closed and P^n is open.

As an example of a set which is neither open nor closed, let

$$X = \{\mathbf{x} \in R^2 : x_1 > 1 \quad \text{and} \quad x_2 \geq 1\}.$$

This set is depicted in Figure 9.1. The continuous horizontal line H, excluding A = (1, 1), is that part of bdry X which is in X. The broken vertical line V, including the point A, is that part of bdry X which is in X^c.

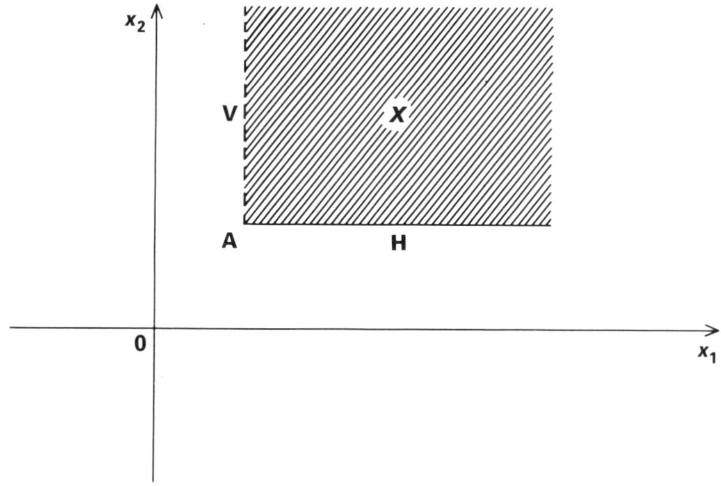

FIG. 9.1

The useful fact about open sets for our purposes is that, given a point $\mathbf{x}$ in such a set, one can move some distance from $\mathbf{x}$ in any direction without passing outside the set. This makes open sets the natural habitat for differentiable functions.

A *function* on a set X is some rule which assigns to each member $\mathbf{x}$ of X a real number $f(\mathbf{x})$. The set X is called the *domain* of the function f. Books on analysis usually emphasise the distinction between a function and the values it takes, illustrating this point with an analogy about (say) ovens and bread. The distinction is, of course, important but the phrase 'the function $f(\mathbf{x})$' is so useful for so many purposes that we shall not hesitate to use it. Notice that a function $f(\mathbf{x})$ defined on a set in R^n is a function of n real variables $x_1, \ldots, x_n$; the fact that we are writing vectors as columns does not affect the truth of this statement.

The function $f(\mathbf{x})$ defined on an open set U in R^n is said to be of *class* C^1 if it and its partial derivatives

$$\partial f/\partial x_1, \ldots, \partial f/\partial x_n$$

are continuous functions of n for all $\mathbf{x}$ in U. Notice that we have slipped in the assumption of continuity as well as existence of partial derivatives: not everything in this chapter requires the full set of C^1 assumptions, but the main theorems do, so we may as well impose C^1 from the start. At one point in the next chapter we shall consider second derivatives: a C^1 function is of class C^2 if its partial derivatives are themselves of class C^1.

Let f be a C^1 function on an open set U and let $\mathbf{a} \in U$. The *gradient* of f at $\mathbf{a}$ is defined to be the n-vector of partial derivatives

$$(\partial f/\partial x_1 \cdots \partial f/\partial x_n)^{\mathrm{T}}$$

evaluated at $\mathbf{x} = \mathbf{a}$. This is denoted by

$$\text{grad}\, f(\mathbf{a}) \quad \text{or} \quad \nabla f(\mathbf{a})$$

where ∇ is pronounced 'dell'.

Listing partial derivatives as a vector adds nothing to what the reader is presumed to know already. What is perhaps less familiar is the notion of gradient as a limit.

Consider first the case where $n = 2$. Let f be of class C^1 on an open set U in R^2, let $\mathbf{a} \in U$ and let $\mathbf{z} = \nabla f(\mathbf{a})$. Let $\mathbf{b}$ be any 2-vector. Since U is open, $\mathbf{a} + \varepsilon\mathbf{b}$ is in U for all sufficiently small ε. Given that f is C^1, we can 'expand' the expression $f(\mathbf{a} + \varepsilon\mathbf{b})$ in terms of ε. Specifically,

$$f(\mathbf{a} + \varepsilon\mathbf{b}) = f(\mathbf{a}) + (\varepsilon b_1)z_1 + (\varepsilon b_2)z_2 + r(\varepsilon),$$

where the remainder term $r(\varepsilon)$ is small relative to ε when ε is small. Dividing by ε and rearranging we have

$$(f(\mathbf{a} + \varepsilon\mathbf{b}) - f(\mathbf{a}))/\varepsilon = b_1 z_1 + b_2 z_2 + r(\varepsilon)/\varepsilon.$$

As ε approaches zero, so does $r(\varepsilon)/\varepsilon$. This and the definition of $\mathbf{z}$ imply that

$$\lim_{\varepsilon \to 0} \left[\frac{f(\mathbf{a} + \varepsilon\mathbf{b}) - f(\mathbf{a})}{\varepsilon}\right] = \mathbf{b}^{\mathrm{T}}\, \nabla f(\mathbf{a}). \tag{2}$$

Now, (2) is the fundamental relation which holds not just for $n = 2$ but for every n. It implies that if

$\mathbf{b}^{\mathrm{T}}\, \nabla f(\mathbf{a}) > 0$ then $f(\mathbf{a} + \varepsilon\mathbf{b}) > f(\mathbf{a})$ for all sufficiently small positive ε.

This result will be used to great effect in the next section.

Exercises

1. Give examples of sets in R^2 which are
   (a) closed and bounded,
   (b) closed but not bounded,
   (c) bounded but not closed,
   (d) closed and open.
2. Show that the intersection of two open sets in R^n is open.
3. On P^3, define $f(\mathbf{x}) = (x_1 + x_2)/(x_2 + x_3)$. Give a formula for $\nabla f(\mathbf{x})$.

9.2 The Fritz John Theorem

This section and the next are concerned with local maxima. To explain what these are, we must first define an 'open neighbourhood'.

Given a point $\mathbf{a}$ of R^n, an *open neighbourhood* of $\mathbf{a}$ is any open set which owns $\mathbf{a}$. Examples of open neighbourhoods of $\mathbf{a}$ are open balls which own $\mathbf{a}$, including those with centre $\mathbf{a}$. Another open neighbourhood of $\mathbf{a}$ is R^n itself: having no boundary points, R^n is open.

Let f be a function on R^n and let $\hat{\mathbf{x}} \in R^n$. We say that f has a *local maximum* at $\hat{\mathbf{x}}$ if there exists an open neighbourhood U of $\hat{\mathbf{x}}$ such that

$$f(\hat{\mathbf{x}}) \geq f(\mathbf{x}) \quad \text{for all } \mathbf{x} \in U.$$

If U can be taken to be all of R^n, we say that f has a *maximum* at $\hat{\mathbf{x}}$.

Notice that a maximum is a local maximum, but a local maximum may not be a maximum. If $n = 1$ and $f(x) = x(x-1)^2$ there is no maximum but there is a local maximum at $x = \frac{1}{3}$: we can take the open set U to be the set of all real numbers which are less than 1.

Mathematicians generally prefer the term 'relative maximum' to 'local maximum'. This terminology loses its attraction when constraints are introduced: ambiguities arise as to whether 'relative' means 'relative to an open neighbourhood' or 'relative to the explicit constraints'. Since we shall mostly be concerned with constrained problems, we use 'local' throughout.

We now state the theorem about local maxima that everybody knows.

THEOREM 1 Let f be a C^1 function on R^n which has a local maximum at $\hat{\mathbf{x}}$. Then $\nabla f(\hat{\mathbf{x}}) = \mathbf{0}$.

PROOF Let $\mathbf{a}$ be a point such that $\nabla f(\mathbf{a})$ is some non-zero vector, say

$\mathbf{z}$. Then $\mathbf{z}^T\mathbf{z} > 0$, so $f(\mathbf{a} + \varepsilon\mathbf{z}) > f(\mathbf{a})$ for all sufficiently small positive ε. Thus f does not have a local maximum at $\mathbf{a}$. QED

One thing to notice about Theorem 1 is that the assumption that f be defined on all of R^n is quite unnecessary: it suffices that f be defined and of class C^1 on some open set V which owns $\hat{\mathbf{x}}$. Why do we insist that V be open? One answer is that it rules out boundary optima: if $n = 1$ and $f(x)$ is x for $x \leq 0$ and is undefined elsewhere, then the maximum occurs at 0 and $f'(0) = 1$. Notice, however, that examples like this violate the basic principle of the last section, which is the other answer to our question: C^1 functions live on open sets.

We now bring in constraints. To avoid messy notation, we shall assume that all relevant functions are defined on all of R^n; but the reader should bear in mind that the theorems remain true when $f_0, f_1, \ldots, f_m$ are defined and C^1 on an open neighbourhood V of the point of interest $\hat{\mathbf{x}}$.

Let f be a function of R^n and let X be a set in R^n. Consider the problem

$$\text{maximise } f(\mathbf{x}) \quad \text{subject to } \mathbf{x} \in X.$$

We say that f has a maximum at $\hat{\mathbf{x}}$ subject to $\mathbf{x} \in X$ if $\hat{\mathbf{x}}$ is a solution vector for this problem. We say that f has a local maximum at $\hat{\mathbf{x}}$ subject to $\mathbf{x} \in X$ if, for some open neighbourhood U of $\hat{\mathbf{x}}$, the problem

$$\text{maximise } f(\mathbf{x}) \quad \text{subject to } \mathbf{x} \in X \cap U$$

has $\hat{\mathbf{x}}$ as a solution vector.

We start by considering the case where all constraints take the form of weak inequalities. We write such constraints in the form $f(\mathbf{x}) \geq 0$. This is quite general, since a constraint like $g(\mathbf{x}) \leq \beta$ can be put in the given form by setting $f(\mathbf{x}) = \beta - g(\mathbf{x})$.

THEOREM 2 Let $f_0, f_1, \ldots, f_m$ be C^1 functions on R^n. Let f_0 have a local maximum at $\hat{\mathbf{x}}$ subject to the constraints

$$f_i(\mathbf{x}) \geq 0 \quad (i = 1, \ldots, m).$$

Then there exist multipliers $\phi_0, \phi_1, \ldots, \phi_m$, all non-negative and not all zero, such that

$$\sum_{i=0}^{m} \phi_i \nabla f_i(\hat{\mathbf{x}}) = 0. \tag{1}$$

PROOF This is a simple application of the Finite Separation Theorem (8.3.2). Let $\mathbf{z}_i = \nabla f_i(\hat{\mathbf{x}})$ for $i = 0, 1, \ldots, m$. It suffices to show that the n-vector $\mathbf{0}$ is a convex combination of $\mathbf{z}_0, \mathbf{z}_1, \ldots, \mathbf{z}_m$.

Suppose the contrary. Then by the Finite Separation Theorem there exists some $\mathbf{p}$ such that $\mathbf{p}^{\mathrm{T}}\mathbf{z}_i > 0$ for $i = 0, 1, \ldots, m$. But then for all sufficiently small positive ε, $f_0(\hat{\mathbf{x}} + \varepsilon\mathbf{p}) > f_0(\hat{\mathbf{x}})$ and

$$f_i(\hat{\mathbf{x}} + \varepsilon\mathbf{p}) > f_i(\hat{\mathbf{x}}) \geq 0 \quad \text{for } i = 1, \ldots, m.$$

Thus we can find points arbitrarily close to $\hat{\mathbf{x}}$ which satisfy the constraints and make f_0 larger than it is at $\hat{\mathbf{x}}$: the local-maximum hypothesis is contradicted. QED

In long hand, (1) states that, at $\mathbf{x} = \hat{\mathbf{x}}$,

$$\phi_0(\partial f_0/\partial x_j) + \phi_1(\partial f_1/\partial x_j) + \cdots + \phi_m(\partial f_m/\partial x_j) = 0$$

for $j = 1, \ldots, n$. If $\phi_0 > 0$ we may set $\lambda_i = \phi_i/\phi_0$ for $i = 1, \ldots, m$. Then $\lambda_1, \ldots, \lambda_m$ are non-negative multipliers such that, at $\mathbf{x} = \hat{\mathbf{x}}$,

$$(\partial f_0/\partial x_j) + \lambda_1(\partial f_1/\partial x_j) + \cdots + \lambda_m(\partial f_m/\partial x_j) = 0 \tag{2}$$

for $j = 1, \ldots, n$. Now, (2) should have a familiar ring to readers accustomed to Lagrange multipliers. But notice that to get (2) from (1) we had to assume that $\phi_0 > 0$, whereas Theorem 2 merely asserts that $\phi_0, \phi_1, \ldots, \phi_m$ are non-negative and not all of them are zero. We shall have much more to say about this in the next section; for the moment we content ourselves with an example which shows that $\phi_0 > 0$ cannot be guaranteed.

Example Let $n = 2$. Suppose we wish to maximise a function $f(\mathbf{x})$ subject to the constraints

$$x_1 \geq 0 \quad \text{and} \quad x_1 + x_2^2 \leq 0.$$

It is easy to see that the only 2-vector $\mathbf{x}$ which satisfies these constraints is $\mathbf{0}$. So, regardless of the form of f, there is a constrained maximum at $\mathbf{0}$.

We may therefore apply Theorem 2 with $m = 2$, $f_0 = f$, $f_1(\mathbf{x}) = x_1$, $f_2(\mathbf{x}) = -x_1 - x_2^2$ and $\hat{\mathbf{x}} = \mathbf{0}$. At any $\mathbf{x}$,

$$\nabla f_1(\mathbf{x}) = \begin{pmatrix} 1 \\ 0 \end{pmatrix} \quad \text{and} \quad \nabla f_2(\mathbf{x}) = \begin{pmatrix} -1 \\ -2x_2 \end{pmatrix}.$$

In particular, $\nabla f_1(\mathbf{0}) = -\nabla f_2(\mathbf{0}) = (1 \;\; 0)^{\mathrm{T}}$. Denote the 2-vector $\nabla f(\mathbf{0})$

by $\mathbf{z}$: by Theorem 2 there exist non-negative multipliers ϕ_0, ϕ_1, ϕ_2, not all zero, such that

$$\phi_0\begin{pmatrix} z_1 \\ z_2 \end{pmatrix} + (\phi_1 - \phi_2)\begin{pmatrix} 1 \\ 0 \end{pmatrix} = \begin{pmatrix} 0 \\ 0 \end{pmatrix}. \tag{3}$$

Clearly (3) is satisfied by $\phi_0 = 0$, $\phi_1 = \phi_2 = 1$: thus we have done nothing to contradict Theorem 2. The question now arises: can we find some other list of ϕ's satisfying the conditions of Theorem 2 and $\phi_0 > 0$? The answer is, in general, no. For the second equation of (3) reads $\phi_0 z_2 = 0$: so if f is any function such that

$$\partial f/\partial x_2 \neq 0 \quad \text{when } x = 0$$

then $\phi_0 = 0$.

We now turn to the case where the constraints take the form of equations as well as weak inequalities. We use a partition notation similar to that of Section 8.1.

THEOREM 3 Let $f_0, f_1, \ldots, f_m$ be C^1 functions on R^n and let $\{M_1, M_2\}$ be a partition of m. Let f_0 have a local maximum at $\hat{\mathbf{x}}$ subject to the constraints

$$f_i(\mathbf{x}) \geq 0 \quad (i \in M_1),$$
$$f_i(\mathbf{x}) = 0 \quad (i \in M_2).$$

Then there exist multipliers $\phi_0, \phi_1, \ldots, \phi_m$, satisfying (1) and not all zero, such that $\phi_i \geq 0$ for $i = 0$ and all i in M_1; ϕ_i can have either sign for i in M_2.

PROOF We give a non-rigorous argument and a rigorous one. We hope that most readers will find the non-rigorous argument instructive. The rigorous argument uses the Weierstrass Maximum Value Theorem mentioned at the end of Section 5.2 (and proved in Bartle, 1976, pp. 154–5) and the Bolzano–Weierstrass Theorem (Bartle, 1976, p. 108) which states that any bounded sequence in R^n has a convergent subsequence. Readers unfamiliar with these theorems can omit the rigorous argument.

Non-rigorous argument Suppose for definiteness that

$$M_1 = \{1, 2, \ldots, k-1\} \quad \text{and} \quad M_2 = \{k, k+1, \ldots, m\}.$$

Let $X = \{\mathbf{x} \in R^n\colon f_i(\mathbf{x}) \geq 0 \text{ for } i = 1, \ldots, m\}$ and define the function

$$g(\mathbf{x}) = \sum_{i=k}^{m} f_i(\mathbf{x}).$$

Then $g(\mathbf{x}) \geq 0$ for all $\mathbf{x}$ in X, with equality if and only if $f_i(\mathbf{x}) = 0$ for all $i \geq k$. Thus the constraints of the problem may be written

$$\mathbf{x} \in X \quad \text{and} \quad g(\mathbf{x}) \leq 0.$$

Notice that the introduction of g to handle the equation constraints is just the same trick we use when setting up an auxiliary programme in linear programming. But instead of solving an auxiliary programme, we pass straight to an application of Theorem 2.

This theorem ensures the existence of multipliers $\mu_0, \mu_1, \ldots, \mu_m$ and μ_*, all non-negative and not all zero, such that

$$\sum_{i=0}^{m} \mu_i \nabla f_i(\hat{\mathbf{x}}) = \mu_* \nabla g(\hat{\mathbf{x}}).$$

Let $\phi_i = \mu_i$ for $0 \leq i < k$, $\phi_i = \mu_i - \mu_*$ for $i \geq k$. Then $\phi_i \geq 0$ for $i < k$ and ϕ_i may have either sign for $i \geq k$. By definition of g, the multipliers $\phi_0, \phi_1, \ldots, \phi_m$ satisfy (1). Assuming that these numbers are not all zero, we have the result.

The reason why this argument is non-rigorous is that the assumption at the end is dubious: we have done nothing to rule out the case where

$$\begin{aligned} &\mu_i = 0 && \text{for } i = 0, 1, \ldots, k-1 \\ \text{and } &\mu_i = \mu_* > 0 && \text{for } i = k, \ldots, m. \end{aligned}$$

This is a serious gap: Theorem 3, *sans* the requirement that the ϕ's be not all zero, reduces to the rather uninteresting proposition that $0 = 0$.

Rigorous argument The proof is a modification of McShane (1973).

Let X and g be as in the non-rigorous argument: notice that X is closed. By definition of a local maximum we may choose an open set U such that $\hat{\mathbf{x}}$ maximises $f_0(\mathbf{x})$ subject to

$$\mathbf{x} \in X \cap U \quad \text{and} \quad g(\mathbf{x}) \leq 0. \tag{4}$$

Define the function

$$f(\mathbf{x}) = f_0(\mathbf{x}) - f_0(\hat{\mathbf{x}}) - \|\mathbf{x} - \hat{\mathbf{x}}\|^2.$$

Then $f(\mathbf{x}) \leq 0$ for all $\mathbf{x}$ satisfying (4), with equality if *and only if* $\mathbf{x} = \hat{\mathbf{x}}$. Also, f and f_0 have the same gradient at $\hat{\mathbf{x}}$: thus it suffices to prove the theorem with f_0 replaced by f.

Let B, $\bar{B}$ and S denote the open ball, closed ball and sphere with centre $\hat{\mathbf{x}}$ and radius δ, where δ is chosen sufficiently small that $\bar{B} \subset U$.

Let $Y = X \cap \bar{B}$ and split this set into three sets as follows:

$$Y_1 = X \cap B,$$
$$Y_2 = \{\mathbf{x} \in X \cap S\colon f(\mathbf{x}) < 0\},$$
$$Y_3 = \{\mathbf{x} \in X \cap S\colon f(\mathbf{x}) \geq 0\}.$$

Now S is contained in U and does not own $\hat{\mathbf{x}}$: thus $g(\mathbf{x}) > 0$ for all $\mathbf{x}$ in Y_3. We may therefore apply Weierstrass's Theorem to the continuous function $f(\mathbf{x})/g(\mathbf{x})$ on the closed and bounded set Y_3. This enables us to choose a positive number λ such that $f(\mathbf{x}) < \lambda g(\mathbf{x})$ for all $\mathbf{x}$ in Y_3; if Y_3 is empty, any positive λ will do. Now

$$f(\mathbf{x}) < 0 \leq g(\mathbf{x}) \quad \text{for all } \mathbf{x} \text{ in } Y_2 .$$

Also $f(\hat{\mathbf{x}}) = 0 = g(\hat{\mathbf{x}})$ and $\hat{\mathbf{x}} \in Y_1$. Thus the function

$$h(\mathbf{x}) = f(\mathbf{x}) - \lambda g(\mathbf{x})$$

satisfies $h(\mathbf{x}) < 0$ for all $\mathbf{x}$ in $Y_2 \cup Y_3$ and $h(\mathbf{x}) \geq 0$ for some $\mathbf{x}$ in Y_1. Again by Weierstrass, we can choose some $\mathbf{x}^*$ which maximises $h(\mathbf{x})$ subject to $\mathbf{x} \in Y$. Then $x^* \in Y_1$, which implies two things. First,

$$\|\mathbf{x}^* - \hat{\mathbf{x}}\| < \delta. \tag{5}$$

Second, $\mathbf{x}^*$ maximises $h(\mathbf{x})$ subject to $\mathbf{x} \in X \cap B$. Since B is open, h has a local maximum at $\mathbf{x}^*$ subject to $\mathbf{x} \in X$.

Apply Theorem 2 to h and $\mathbf{x}^*$: there exist non-negative multipliers $\alpha_0, \alpha_1, \ldots, \alpha_m$, all non-negative and not all zero, such that

$$\alpha_0 \nabla h(\mathbf{x}^*) + \sum_{i=1}^{m} \alpha_i \nabla f_i(\mathbf{x}^*) = \mathbf{0}.$$

By definition of h and g, this equation may be written

$$\beta_0 \nabla f(\mathbf{x}^*) + \sum_{i=1}^{m} \beta_i \nabla f_i(\mathbf{x}^*) = \mathbf{0} \tag{6}$$

where $\beta_i = \alpha_i$ for $i < k$ and $\beta_i = \alpha_i - \lambda\alpha_0$ for $i \geq k$. Since the α's are not all zero, the β's are not all zero. Also

$$\beta_i \geq 0 \quad \text{for } i = 0, 1, \ldots, k-1. \tag{7}$$

To summarise: for all sufficiently small δ there exist a point $\mathbf{x}^*$ and multipliers $\beta_0, \beta_1, \ldots, \beta_m$ which are not all zero, satisfying (5)–(7) inclusive. Notice that $\mathbf{x}^*$ and the β's depend on δ. Since $\beta_0, \beta_1, \ldots, \beta_m$ retain their properties when multiplied by a positive scalar, we may scale them so that

$$\sum_{i=0}^{m} |\beta_i| = 1. \tag{8}$$

Now let δ approach zero through a sequence of values. By (5), $\mathbf{x}^*$ approaches $\hat{\mathbf{x}}$. By (8) and Bolzano–Weierstrass Theorem, the sequence may be chosen in such a way that $\beta_0, \beta_1, \ldots, \beta_m$ approach limits $\phi_0, \phi_1, \ldots, \phi_m$. Since all our functions are of class C^1, (6)–(8) hold with $\mathbf{x}^*$ replaced by $\hat{\mathbf{x}}$ and the β's by the ϕ's. QED

Now for a nice simple result about complementarity. Let notation and assumptions be as in Theorem 3. We split M_1 into two sets M_{1s} and M_{1t}, where s stands for 'slack' and t for 'tight'. Specifically, M_{1s} consists of those i in M_1 for which $f_i(\hat{\mathbf{x}}) > 0$ and M_{1t} of those i in M_1 for which $f_i(\hat{\mathbf{x}}) = 0$. Notice the distinction between M_2 and M_{1t}: the statement

$$f_i(\mathbf{x}) = 0 \quad (i \in M_2)$$

is a constraint of the problem, whereas the statement

$$f_i(\hat{\mathbf{x}}) = 0 \quad (i \in M_{1t})$$

is a property of the (local) solution $\hat{\mathbf{x}}$.

The complementarity result is that the multipliers $\phi_0, \phi_1, \ldots, \phi_m$ of Theorem 3 can be chosen so that

$$\phi_i = 0 \quad \text{for all } i \text{ in } M_{1s}.$$

The proof is easy. For all i in M_{1s} we have $f_i(\hat{\mathbf{x}}) > 0$. By continuity, $f_i(\mathbf{x}) > 0$ for all such i and all $\mathbf{x}$ sufficiently close to $\hat{\mathbf{x}}$. Since we are interested only in *local* maxima we can apply Theorem 3 with the M_{1s} constraints ignored.

We can summarise Theorem 3 and the last result as follows.

THEOREM 4 (Fritz John Theorem)
Let $f_0, f_1, \ldots, f_m$ be C^1 functions on R^n and let $\{M_1, M_2\}$ be a partition of m. Let f_0 have a local maximum at $\hat{\mathbf{x}}$ subject to the constraints

$$f_i(\mathbf{x}) \geq 0 \quad (i \in M_1),$$
$$f_i(\mathbf{x}) = 0 \quad (i \in M_2).$$

Let $M_1 = M_{1s} \cup M_{1t}$ where

$$f_i(\hat{\mathbf{x}}) > 0 \quad (i \in M_{1s}),$$
$$f_i(\hat{\mathbf{x}}) = 0 \quad (i \in M_{1t}).$$

Then there exist multipliers $\phi_0, \phi_1, \ldots, \phi_m$ with the following properties:

(FJ1) $\sum_{i=0}^{m} \phi_i \nabla f_i(\hat{\mathbf{x}}) = \mathbf{0}$,

(FJ2) $\phi_i = 0$ for all i in M_{1s},

(FJ3) $\phi_i \geq 0$ for all i in M_{1t},

(FJ4) $\phi_0 \geq 0$ and $\phi_0, \phi_1, \ldots, \phi_m$ are not all zero.

For i in M_2, ϕ_i may have either sign.

Notice that Theorem 4 includes Theorem 3. Also, Theorem 2 is the special case of Theorem 3 when M_2 is empty.

9.3 The Kuhn–Tucker Theorem

There were two main results in the last section. One was the Fritz John Theorem. The other was the nasty little example which shows that $\phi_0 > 0$ cannot be guaranteed.

In a sense, these results cancel each other out. The Fritz John Theorem gives necessary conditions for a local constrained maximum. The only place where the objective function f_0 appears in these conditions is in (FJ1), where the gradient of f_0 is multiplied by the non-negative number ϕ_0. If we cannot be sure that $\phi_0 > 0$ we cannot be sure that the necessary conditions involve the objective function at all, so we can hardly expect them to tell us very much about its local maxima. And the example showed that it is not always the case that $\phi_0 > 0$.

The way out of this quandary is to impose more assumptions. Conditions on the constraints that ensure the existence of multipliers such that $\phi_0 > 0$ are called *constraint qualifications*.

Let notation be as in Theorem 9.2.4 and let

$$\mathbf{z}_i = \nabla f_i(\hat{\mathbf{x}}) \quad \text{for } i = 1, \ldots, m.$$

We can derive one constraint qualification straight away by looking at (FJ1) and (FJ4). If $\phi_0 = 0$, then

$$\phi_1 \mathbf{z}_1 + \cdots + \phi_m \mathbf{z}_m = \mathbf{0}$$

by (FJ1), and $\phi_1, \ldots, \phi_m$ are not all zero by (FJ4). Thus $\phi_0 = 0$ implies the linear *dependence* of the m n-vectors $\mathbf{z}_1, \ldots, \mathbf{z}_n$. Equivalently, linear *independence* of these vectors is a constraint qualification.

To obtain a slightly more general constraint qualification we introduce the set $\hat{M}$ of all those $i = 1, \ldots, m$ for which $f_i(\hat{\mathbf{x}}) = 0$. Thus

$$\hat{M} = M_{1t} \cup M_2 = \{i \geq 1 \colon i \text{ is not in } M_{1s}\}.$$

Repeating the argument of the preceding paragraph, but using (FJ2) as well as (FJ1) and (FJ4), we see that $\phi_0 = 0$ implies linear dependence of the vectors $\{\mathbf{z}_i \colon i \in \hat{M}\}$. So linear independence of these vectors is a constraint qualification.

If a constraint qualification is satisfied we may set $\lambda_i = \phi_i / \phi_0$ for $i = 1, \ldots, m$: then (FJ1)–(FJ3) remain true when ϕ_i is replaced by λ_i for $i \geq 1$ and ϕ_0 is replaced by 1. This yields the following theorem.

THEOREM 1 (Kuhn–Tucker Theorem)
Let assumptions and notation be as in the Fritz John Theorem and let $\hat{M} = M_{1t} \cup M_2$. Suppose also that the vectors $\{\nabla f_i(\hat{\mathbf{x}}) \colon i \in \hat{M}\}$ are linearly independent.

Then there exist multipliers $\lambda_1, \ldots, \lambda_m$ such that:

$$\text{(KT1)} \quad \nabla f_0(\hat{\mathbf{x}}) + \sum_{i=1}^{m} \lambda_i \nabla f_i(\hat{\mathbf{x}}) = \mathbf{0},$$

$$\text{(KT2)} \quad \lambda_i = 0 \quad \text{for all } i \text{ in } M_{1s},$$

$$\text{(KT3)} \quad \lambda_i \geq 0 \quad \text{for all } i \text{ in } M_{1t}.$$

For i in M_2, λ_i may have either sign.

The following example taken from the theory of the rational consumer illustrates the use of $\hat{M}$ in the Kuhn–Tucker Theorem.

Example 1 We wish to maximise a utility function $u(x_1, x_2)$ subject to the budget constraint $p_1 x_1 + p_2 x_2 \leq y$ and the non-negativity constraints $x_1 \geq 0$ and $x_2 \geq 0$. Here p_1, p_2 and y are positive numbers (prices and income).

We write the constraints in the form

$$f_i(\mathbf{x}) \geq 0 \quad \text{for } i = 1, 2, 3$$

by setting $f_1(\mathbf{x}) = x_1$, $f_2(\mathbf{x}) = x_2$, $f_3(\mathbf{x}) = y - p_1x_1 - p_2x_2$.

At any $\mathbf{x}$ the gradients of these functions are

$$\nabla f_i(\mathbf{x}) = \begin{pmatrix} 1 \\ 0 \end{pmatrix}, \quad \nabla f_2(\mathbf{x}) = \begin{pmatrix} 0 \\ 1 \end{pmatrix}, \quad \nabla f_3(\mathbf{x}) = \begin{pmatrix} -p_1 \\ -p_2 \end{pmatrix}.$$

These three 2-vectors are linearly dependent. But notice that p_1, p_2 and y are positive, so *at most two* of the constraints can hold with equality at any point, and in particular at any local maximum point. Also *any two* of the three gradients are linearly independent. Thus the conditions of the Kuhn–Tucker Theorem are satisfied. We may therefore choose multipliers λ_1, λ_2, λ_3 satisfying (KT1)–(KT3).

Setting $\mu = \lambda_3$ we can write the necessary conditions for a local maximum in the following concise form:

$$\begin{aligned} &\partial u/\partial x_1 \leq \mu p_1 \quad \text{with equality if } x_1 > 0, \\ &\partial u/\partial x_2 \leq \mu p_2 \quad \text{with equality if } x_2 > 0, \\ &\mu \geq 0 \quad \text{with equality if } p_1x_1 + p_2x_2 < y. \end{aligned}$$

The number μ, evaluated at the optimum, is called the *marginal utility of income*; later, we shall see why.

We could have arrived at the results of Example 1 by a different route. All the constraints there are linear, and *linearity of all constraints is a constraint qualification.* This fact is the content of the next theorem, which uses complementarity notation explicitly. The proof does not invoke the Fritz John Theorem, but this does not mean that we are taking a totally different tack. The Fritz John Theorem depends on Theorem 9.2.1, which depends on the Finite Separation Theorem (8.2.2), which depends on Farkas' Lemma (Theorem 8.2.1). Here we use Farkas' Lemma directly.

THEOREM 2 Let $\mathbf{G}$ be an $m \times n$ matrix, $\mathbf{b}$ an m-vector, f a C^1 function on R^n. Let f have a local maximum at $\hat{\mathbf{x}}$ subject to the constraints $\mathbf{Gx} \geq \mathbf{b}$.

Then there exists a non-negative m-vector $\mathbf{p}$ such that

$$\nabla f(\hat{\mathbf{x}}) + \mathbf{G}^{\mathrm{T}}\mathbf{p} = \mathbf{0} \quad \text{and} \quad \mathbf{p} /\!/ \mathbf{G}\hat{\mathbf{x}} - \mathbf{b}.$$

PROOF Let $\mathbf{y} = \mathbf{G}\hat{\mathbf{x}} - \mathbf{b}$, $\mathbf{z} = \nabla f(\hat{\mathbf{x}})$. Suppose for definiteness that the first r components of $\mathbf{y}$ are zero and the rest are positive. We want to express $-\mathbf{z}$ as a linear combination, with non-negative weights, of the first r columns of $\mathbf{G}^{\mathrm{T}}$.

Suppose this cannot be done. By Farkas' Lemma, there exists an n-vector $\mathbf{q}$ such that the first r components of $\mathbf{Gq}$ are non-negative and $\mathbf{q}^{\mathrm{T}}\mathbf{z} > 0$. Let $\mathbf{w} = \mathbf{Gq}$. For $i \leq r$, $y_i = 0$ and $w_i \geq 0$. For $i > r$, $y_i > 0$ and w_i may have either sign. So for all sufficiently small positive ε, $\mathbf{y} + \varepsilon\mathbf{w} \geq \mathbf{0}$; in other words, $\mathbf{G}(\mathbf{x} + \varepsilon\mathbf{q}) \geq \mathbf{b}$. But since $\mathbf{q}^{\mathrm{T}}\mathbf{z} > 0$ we have $f(\hat{\mathbf{x}} + \varepsilon\mathbf{q}) > f(\hat{\mathbf{x}})$ for all sufficiently small positive ε. These inequalities contradict the assumption that f has a local maximum at $\hat{\mathbf{x}}$ subject to the constraints. QED

Using Theorem 2 we can establish a link between the Kuhn–Tucker Theorem and duality in linear programming. Let the n-vector $\hat{\mathbf{x}}$ be a solution vector for the familiar problem

(S) maximise $\mathbf{c}^{\mathrm{T}}\mathbf{x}$ subject to $\mathbf{Ax} \leq \mathbf{b}$ and $\mathbf{x} \geq \mathbf{0}$.

Applying Theorem 2 with due attention to the directions of the inequalities, we obtain a non-negative $(m + n)$-vector $(\mathbf{p} : \mathbf{q})$ such that

$$\mathbf{c} - \mathbf{A}^{\mathrm{T}}\mathbf{p} + \mathbf{q} = \mathbf{0} \quad \text{and} \quad (\mathbf{p} : \mathbf{q}) // (\mathbf{b} - \mathbf{A}\hat{\mathbf{x}} : \hat{\mathbf{x}}).$$

Eliminating $\mathbf{q}$, we see that $(\hat{\mathbf{x}} : \mathbf{b} - \mathbf{A}\hat{\mathbf{x}})$ and $(\mathbf{A}^{\mathrm{T}}\mathbf{p} - \mathbf{c} : \mathbf{p})$ are non-negative and complementary. Given that $\hat{\mathbf{x}}$ is a solution vector for (S), this is equivalent to the statement that $\mathbf{p}$ is a solution vector for the dual (D). Thus the set of solutions to (D) is precisely the set of m-vectors which can serve as multipliers associated with the constraints $\mathbf{Ax} \leq \mathbf{b}$.

The fact that dual solutions in linear programming are special cases of Kuhn–Tucker multipliers suggests the question: can multipliers in non-linear problems be given a marginal interpretation as in Section 7.3? The answer is 'sometimes'.

The story runs as follows. Let f, $a_1, \ldots, a_m$ be C^1 functions on R^n and let $\mathbf{b}$ be an m-vector. Let us assume

(I) the problem, maximise $f(\mathbf{x})$ subject to

$$a_i(\mathbf{x}) \leq b_i \quad \text{for } i = 1, \ldots, m,$$

has a unique solution $\hat{\mathbf{x}}$.

Suppose for definiteness that

$$a_i(\hat{\mathbf{x}}) = b_i \quad \text{for } i = 1, \ldots, r \qquad (1)$$

and that $a_i(\hat{\mathbf{x}}) < b_i$ for $i = r + 1, \ldots, m$. Assume that a constraint qualification is satisfied at $\hat{\mathbf{x}}$ and let $\mathbf{z}_i = \nabla a_i(\hat{\mathbf{x}})$ for $i = 1, \ldots, m$. By Theorem 1, with $f_i(\mathbf{x}) = b_i - a_i(\mathbf{x})$ for $i \geq 1$, there exist non-negative multipliers $\lambda_1, \ldots, \lambda_m$ such that

$$\nabla f(\hat{\mathbf{x}}) = \sum_{i=1}^{r} \lambda_i \mathbf{z}_i, \tag{2}$$

$$\lambda_i = 0 \quad \text{for } i > r. \tag{3}$$

Notice that the solution vector $\hat{\mathbf{x}}$, the solution value $f(\hat{\mathbf{x}})$, the integer r and the multipliers all depend on $\mathbf{b}$.

Now let $\mathbf{b}$ vary. We can regard the solution value as a function of $\mathbf{b}$, say $v(\mathbf{b})$: this is known as the *valuation function* of the problem. We assume

(II) the components of $\hat{\mathbf{x}}$ are C^1 functions of $\mathbf{b}$.

Thus v is also a C^1 function. Fix some k with $1 \leq k \leq m$ and let

$$\theta_{jk} = \partial \hat{x}_j / \partial b_k \quad \text{for } j = 1, \ldots, n.$$

Evaluating $\partial v / \partial b_k$ by the chain rule and applying (2) we have

$$\partial v / \partial b_k = \sum_{i=1}^{r} \sum_{j=1}^{n} \lambda_i z_{ij} \theta_{jk}.$$

But by (1) and the chain rule,

$$\sum_{j=1}^{n} z_{ij} \theta_{jk} = \delta_{ik} \quad \text{for } i = 1, \ldots, r,$$

where δ_{ik} is 1 if $i = k$ and 0 otherwise. Hence

$$\partial v / \partial b_k = \sum_{i=1}^{r} \lambda_i \delta_{ik}. \tag{4}$$

By definition of the δ's, the right-hand side of (4) is λ_k if $k \leq r$ and zero otherwise. But by (3), $\lambda_k = 0$ if $k > r$. Hence

$$\partial v / \partial b_k = \lambda_k \quad \text{for } k = 1, \ldots, r. \tag{5}$$

Now (5) is the 'required result', in that it gives the multipliers as partial derivatives of the valuation function. In particular, it explains the term 'marginal utility of income' in Example 1. The reason why the answer to our question was only 'sometimes' is that the assumptions (I) and (II) are really rather strong. The assumptions can in some cases be relaxed, *as can the conclusions*: recall that in linear programming it is perfectly all right to talk about a marginal interpretation without requiring differentiability of the valuation function, provided we are careful to distinguish between upward and downward marginal profitabilities. The same sort of thing can be done in the case of concave programming: we shall discuss concave programming in Chapter 11 but not that particular point, which is rather tricky. The general moral is that marginal interpretations of multipliers should be handled with care.

We turn at last to the 'classical case' where all constraints are equations. Nothing really new needs to be said about this, since the Fritz John and Kuhn–Tucker Theorems work just as well when M_1 is empty as when it is not. Contrary to remarks sometimes made in the economics literature (for example Chiang, 1974, p. 704), all-constraints-equations is not itself a constraint qualification. This point is illustrated in Example 2 below.

THEOREM 3 (Lagrange Multiplier Theorem)
Let $f_0, f_1, \ldots, f_m$ be C^1 functions on R^n. Let f_0 have a local maximum at $\hat{\mathbf{x}}$ subject to the constraints

$$f_i(\mathbf{x}) = 0 \quad \text{for } i = 1, \ldots, m.$$

For each $i = 0, 1, \ldots, m$ let $\mathbf{z}_i = \nabla f_i(\hat{\mathbf{x}})$.

If $\mathbf{z}_1, \ldots, \mathbf{z}_m$ are linearly independent, there exist multipliers $\mu_1, \ldots, \mu_m$ such that

$$\mathbf{z}_0 = \mu_1 \mathbf{z}_1 + \cdots + \mu_m \mathbf{z}_m .$$

PROOF Apply Theorem 1 with M_1 empty and let $\mu_i = -\lambda_i$ for $i \geq 1$. QED

Example 2 We alter the example of the last section by replacing inequalities with equations. Let $f(x_1, x_2)$ be a C^1 function such that

$$\partial f / \partial x_2 \neq 0 \quad \text{at } x_1 = x_2 = 0.$$

Suppose we wish to maximise f subject to

$$x_1 = 0 \quad \text{and} \quad x_1 + x_2^2 = 0.$$

Again the only feasible point is the origin, so the maximum occurs there. Applying the Fritz John Theorem and arguing exactly as in the earlier example we have $\phi_0 = 0$. Hence there are no Lagrange multipliers μ_1, μ_2.

We end this section, as we began it, with some remarks about constraint qualifications. So far we have encountered two types: linear independence of constraint gradients in Theorems 1 and 3, linearity of all constraints in Theorem 2. These are not the only known constraint qualifications: we shall encounter some more in Chapter 11 which are specially tailored for the concave case. There is also the constraint qualification introduced by Kuhn and Tucker in their original article (1951). We give a brief and informal description of this.

Consider the following variant of Example 2:

$$\text{maximise } x_2 \quad \text{subject to } x_1 \geq 0 \quad \text{and} \quad x_1 + x_2^3 \leq 0.$$

Here the set of feasible $\mathbf{x}$ no longer degenerates to the origin but the maximum is again at the origin. Applying the Fritz John Theorem as usual, we have $\phi_0 = 0$. What has happened in this example is that linear dependence of the constraint gradients and nonlinearity of the constraints have combined to produce a *cusp* of the feasible region at the optimum. The Kuhn–Tucker constraint qualification is designed specifically to rule out such cusps: a formal statement in n dimensions is rather complicated, and we refer the reader to p. 483 of the cited article for the details.

Exercises

1. Let $\hat{\mathbf{x}}$ maximise $f(\mathbf{x})$ subject to $g(\mathbf{x}) = 0$. Suppose that both functions are C^1 and $\nabla g(\hat{\mathbf{x}}) \neq \mathbf{0}$. By Theorem 3, there exists a multiplier μ such that $\nabla h(\hat{\mathbf{x}}) = \mathbf{0}$, where $h(\mathbf{x}) = f(\mathbf{x}) - \mu g(\mathbf{x})$.

   This does *not* imply that h has a local maximum at $\hat{\mathbf{x}}$. Illustrate this by considering the case where $n = 2$ and

   $$f(x_1, x_2) = x_1 x_2, \quad g(x_1, x_2) = x_1 + x_2 - 2.$$

2. Let $\mathbf{G}$ be an $m \times n$ matrix, $\mathbf{b}$ an m-vector, f a C^1 function on R^n. Let f have a local maximum at $\hat{\mathbf{x}}$ subject to $\mathbf{Gx} = \mathbf{b}$. Then there exists an m-vector $\mathbf{p}$ such that

   $$\nabla f(\hat{\mathbf{x}}) + \mathbf{G}^{\mathrm{T}}\mathbf{p} = \mathbf{0}.$$

   This Lagrange multiplier theorem for linear equation constraints may be proved using either Theorem 2 or Theorem 3. Provide both proofs.

10 Concave Functions

Chapter 9 was about *necessary* conditions for a *local maximum*. This is in stark contrast to Chapters 5–8 which were about *necessary and sufficient* conditions for a *maximum*. Of course, the reason why we were able to establish such strong results in linear programming is that we imposed restrictions on the type of functions we used. This is a general principle, but we do not have to impose quite such drastic restrictions as in linear programming to get interesting, necessary and sufficient conditions for maxima. The appropriate restriction is to require that functions be concave, and that it is what we shall do in Chapter 11.

In this chapter we take a break from maximisation in order to introduce concave functions. We begin in Section 10.1 by making some remarks about convex sets, which are where concave functions live: this will also enable us to give a geometrical interpretation of the Finite Separation Theorem.

Section 10.2 introduces the objects of major interest: our prime purpose is to develop the reader's intuition about what kinds of function are concave. Section 10.3 is about quasi-concave functions, which are more general than concave functions. Quasi-concave functions are important in their own right in economics (production functions, utility functions), but our main object is to show how concave functions can be derived from quasi-concave ones using homogeneity.

10.1 Convex sets

Recall from Section 1.4 that, if $\mathbf{x}$ and $\mathbf{y}$ are n-vectors, the line segment joining them is the set of all points of the form

$$(1-\theta)\mathbf{x}+\theta\mathbf{y}$$

where $0 \leq \theta \leq 1$. We say that a set X in R^n is *convex* if, for any $\mathbf{x}$ and $\mathbf{y}$ in X, the line segment joining $\mathbf{x}$ and $\mathbf{y}$ is contained in X. Figure 10.1

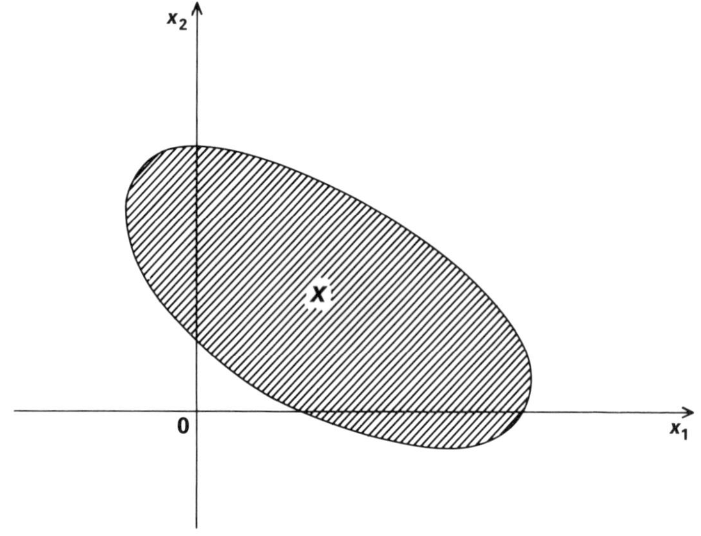

FIG. 10.1 The set X is convex

gives an example of a convex set X in R^2, and Figure 10.2 a non-convex set Y in R^2.

If $\mathbf{x}$ and $\mathbf{y}$ belong to a set X in R^n and θ is either 0 or 1, then

$$(1-\theta)\mathbf{x}+\theta\mathbf{y} \in X$$

whether or not X is convex. Thus the line-segment definition of a convex set is equivalent to the one given in Section 5.3: X is convex if and only if

$$\alpha\mathbf{x}+(1-\alpha)\mathbf{y} \in X$$

whenever $\mathbf{x} \in X$, $\mathbf{y} \in X$ and $0 < \alpha < 1$.

Notice that a set which owns less than two points must, by definition, be convex. Thus $\varnothing$ and all singletons are convex sets.

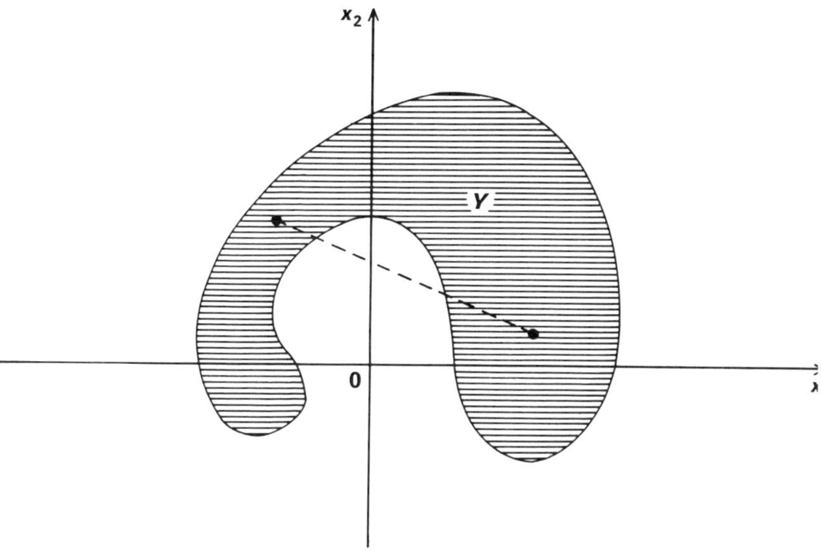

FIG. 10.2 The set Y is not convex

To give more interesting examples of convex sets we start with the case $n = 1$. Of course, R^1 is the set of all real numbers, and we denote it henceforth by R. The set R is sometimes called the *real line* (think of it as the x-axis).

A convex set in R is called an *interval*. There are eleven different kinds of interval. The trivial ones are the empty set (1), the singletons (2) and R itself (3). An interval which is a proper subset of R but owns arbitrarily large positive numbers is said to be bounded below but not above: such an interval may (4) or may not (5) own its left endpoint. Similarly, an interval which is bounded above but not below may (6) or may not (7) own its right endpoint. An interval with two endpoints may own both (8) or neither (9) or only the left (10) or only the right (11). Some authors (for example Bartle, 1976, p. 47) bring the count down to ten by regarding (2) as a special case of (8).

Now consider R^n. Again, the empty set, all singletons and R^n itself are convex. Any linear subspace of R^n is convex. An open ball is convex: if $\|\mathbf{x} - \mathbf{b}\| < \delta$ and $\|\mathbf{y} - \mathbf{b}\| < \delta$ and $0 < \alpha < 1$, then

$$\begin{aligned} \|\alpha\mathbf{x} + (1 - \alpha)\mathbf{y} - \mathbf{b}\| &= \|\alpha(\mathbf{x} - \mathbf{b}) + (1 - \alpha)(\mathbf{y} - \mathbf{b})\| \\ &\leq \|\alpha(\mathbf{x} - \mathbf{b})\| + \|(1 - \alpha)(\mathbf{y} - \mathbf{b})\| \\ &= \alpha\|\mathbf{x} - \mathbf{b}\| + (1 - \alpha)\|\mathbf{y} - \mathbf{b}\| \\ &< \alpha\delta + (1 - \alpha)\delta \\ &= \delta. \end{aligned}$$

By a similar argument, a closed ball is convex.

A *hyperplane* in R^n is the set of the form

$$H = \{\mathbf{x} \in R^n \colon \mathbf{p}^{\mathrm{T}}\mathbf{x} = \beta\}$$

where $\mathbf{p}$ is a non-zero n-vector and β is a scalar. If $\beta = 0$ then H is the set of n-vectors which are orthogonal to $\mathbf{p}$: thus a hyperplane owning $\mathbf{0}$ is the same as an $(n-1)$-dimensional linear subspace. A hyperplane in R^2 is a line, not necessarily through the origin.

The hyperplane H is the boundary of each of the following four sets:

$$H_1 = \{\mathbf{x} \in R^n \colon \mathbf{p}^{\mathrm{T}}\mathbf{x} \leq \beta\}, \quad H_2 = \{\mathbf{x} \in R^n \colon \mathbf{p}^{\mathrm{T}}\mathbf{x} \geq \beta\},$$

$$H_{11} = \{\mathbf{x} \in R^n \colon \mathbf{p}^{\mathrm{T}}\mathbf{x} < \beta\}, \; H_{22} = \{\mathbf{x} \in R^n \colon \mathbf{p}^{\mathrm{T}}\mathbf{x} > \beta\}.$$

The sets H_1 and H_2 are called *closed half-spaces*: they are convex and closed. The sets H_{11} and H_{22} are called *open half-spaces*: they are convex and open.

We can generate other convex sets by using the following theorem.

THEOREM 1 If X and Y are convex sets in R^n then so is $X \cap Y$.

PROOF If $\mathbf{a}$ and $\mathbf{b}$ belong to $X \cap Y$ then the line segment joining $\mathbf{a}$ and $\mathbf{b}$ is contained in X by convexity of X and in Y by convexity of Y and is therefore a subset of $X \cap Y$. QED

Notice that the result of Theorem 1 may be iterated: if X_1, X_2, X_3, X_4 are convex sets in R^n, then so is $X_1 \cap X_2 \cap X_3 \cap X_4$.

As an example of the use of Theorem 1, consider the set

$$X = \{\mathbf{x} \in R^4 \colon x_1 + 2x_2 = x_3 + 3x_4 = 0 \text{ and } x_2 - 5x_3 > 1 \text{ and } \|\mathbf{x}\| \leq 8\}.$$

Then X is the intersection of a linear subspace, an open half-space and a closed ball, so X is convex.

As a second and a third example, consider the positive orthant P^n and the non-negative orthant R^n_+. Both sets are convex, since the former is the intersection of n open half-spaces and the latter the intersection of n closed half-spaces.

A fourth example brings us back to linear programming. Let X be the set of all vectors which are feasible for a given standard-form linear programme. In Section 5.3 we showed directly that such a set is convex. The same result follows from Theorem 1 and the fact that X is an intersection of closed half-spaces.

Another use of the word 'convex' that has already occurred in this book is 'convex combination'. Recall from Section 8.3 that a convex combination of the k n-vectors $\mathbf{x}_1, \ldots, \mathbf{x}_k$ is a vector of the form

$$\alpha_1 \mathbf{x}_1 + \cdots + \alpha_k \mathbf{x}_k$$

where $\alpha_1, \ldots, \alpha_k$ are non-negative scalars whose sum is 1. Given any set X in R^n, finite or infinite, we define a convex combination of members of X to be a convex combination of some finite list of members of X. The set of all convex combinations of members of X is called the *convex hull* of X and is denoted by conv X. The main properties of convex hulls are as follows.

THEOREM 2 If $X \subset R^n$ then
(i) $X \subset \operatorname{conv} X$,
(ii) conv X is a convex set,
(iii) conv $X \subset Y$ for any convex set Y which contains X.

PROOF (i) is obvious, since $\mathbf{x} = 1\mathbf{x}$.

To prove (ii), let

$$\mathbf{x} = \alpha_1 \mathbf{x}_1 + \cdots + \alpha_k \mathbf{x}_k, \quad \mathbf{y} = \beta_1 \mathbf{y}_1 + \cdots + \beta_l \mathbf{y}_l$$

where $\mathbf{x}_1, \ldots, \mathbf{x}_k, \mathbf{y}_1, \ldots, \mathbf{y}_l$ are members of X and $\alpha_1, \ldots, \alpha_k, \beta_1, \ldots, \beta_l$ are non-negative numbers such that

$$\alpha_1 + \cdots + \alpha_k = 1 = \beta_1 + \cdots + \beta_l.$$

Let $\mathbf{z} = (1 - \theta)\mathbf{x} + \theta\mathbf{y}$ where $0 \leq \theta \leq 1$. Set

$$\gamma_i = (1 - \theta)\alpha_i \quad \text{for } i = 1, \ldots, k, \quad \delta_j = \theta\beta_j \quad \text{for } j = 1, \ldots, l.$$

Then $\gamma_1, \ldots, \gamma_k, \delta_1, \ldots, \delta_l$ are $k + l$ non-negative numbers which sum to 1. Also

$$\mathbf{z} = \gamma_1 \mathbf{x}_1 + \cdots + \gamma_k \mathbf{x}_k + \delta_1 \mathbf{y}_1 + \cdots + \delta_l \mathbf{y}_l$$

so $\mathbf{z} \in$ conv X.

To prove (iii), let $X \subset Y$, where Y is convex, and let

$$\mathbf{x} = \alpha_1 \mathbf{x}_1 + \cdots + \alpha_k \mathbf{x}_k$$

where $\alpha_1, \ldots, \alpha_k$ are non-negative numbers which sum to 1. We want to show that $\mathbf{x} \in Y$; this is done by induction on k. The result is immediate if k is 1 or 2. Now suppose that $k = l > 2$ and that the result holds for $k < l$. If $\alpha_1 = 1$ then $\mathbf{x} = \mathbf{x}_1 \in Y$. If $\alpha_1 < 1$ then

$$\mathbf{x} = \alpha_1 \mathbf{x}_1 + (1 - \alpha_1)\mathbf{y} \quad \text{where} \quad \mathbf{y} = (1 - \alpha)^{-1}(\alpha_2 \mathbf{x}_2 + \cdots + \alpha_l \mathbf{x}_l).$$

Then $\mathbf{y} \in Y$ by the induction hypothesis, so $\mathbf{x} \in Y$ by convexity of Y.

QED

Theorem 2 states that the convex hull of a set X is the smallest convex set which contains X. In particular, conv $X = X$ if X is convex.

The convex hull of a finite set in R^n is called a *convex polytope.* Convex polytopes in R^2 are easily illustrated. Let $\mathbf{x}_1$ and $\mathbf{x}_2$ be two different 2-vectors and let $A = \text{conv}\,\{\mathbf{x}_1, \mathbf{x}_2\}$; then A is the line segment joining $\mathbf{x}_1$ and $\mathbf{x}_2$. Let $\mathbf{x}_3$ be a 2-vector which is not in A and let $B = \text{conv}\,\{\mathbf{x}_1, \mathbf{x}_2, \mathbf{x}_3\}$; then B is the set of points on or inside the triangle with vertices $\mathbf{x}_1, \mathbf{x}_2, \mathbf{x}_3$. Let $\mathbf{x}_4$ be a 2-vector which is not in B and let $C = \text{conv}\,\{\mathbf{x}_1, \mathbf{x}_2, \mathbf{x}_3, \mathbf{x}_4\}$: then C is the set of points on or inside the quadrilateral with vertices $\mathbf{x}_1, \mathbf{x}_2, \mathbf{x}_3, \mathbf{x}_4$. And so on.

We can now give the geometric interpretation of the Finite Separation Theorem, 8.3.2. Let $\mathbf{x}_1, \ldots, \mathbf{x}_k$ be points in R^n and let $\mathbf{x}_0$ be a point which is not a convex combination of these. By Theorem 8.3.2, we can choose an n-vector $\mathbf{p}$ such that

$$\mathbf{p}^{\mathrm{T}}\mathbf{x}_0 < \mathbf{p}^{\mathrm{T}}\mathbf{x}_i \quad \text{for } i = 1, \ldots, k.$$

Suppose, for definiteness, that $\mathbf{p}^{\mathrm{T}}\mathbf{x}_1$ is the least of the numbers $\mathbf{p}^{\mathrm{T}}\mathbf{x}_1, \ldots, \mathbf{p}^{\mathrm{T}}\mathbf{x}_k$ and let β be a scalar such that $\mathbf{p}^{\mathrm{T}}\mathbf{x}_0 < \beta < \mathbf{p}^{\mathrm{T}}\mathbf{x}_1$. Then

$$\mathbf{p}^{\mathrm{T}}(\alpha_1\mathbf{x}_1 + \cdots + \alpha_k\mathbf{x}_k) \geq \mathbf{p}^{\mathrm{T}}\mathbf{x}_1 > \beta$$

whenever $\alpha_1, \ldots, \alpha_k$ are non-negative numbers whose sum is 1. Thus

$$\mathbf{p}^{\mathrm{T}}\mathbf{x}_0 < \beta < \mathbf{p}^{\mathrm{T}}\mathbf{x} \quad \text{for all } \mathbf{x} \text{ in } X, \qquad (*)$$

where $X = \text{conv}\,\{\mathbf{x}_1, \ldots, \mathbf{x}_k\}$.

So the Finite Separation Theorem states that if X is a convex polytope in R^n and $\mathbf{x}_0$ is a point not in X, there exist a vector $\mathbf{p}$ and scalar β satisfying $(*)$. This means that the hyperplane

$$H = \{\mathbf{x} \in R^n \colon \mathbf{p}^{\mathrm{T}}\mathbf{x} = \beta\}$$

separates $\mathbf{x}_0$ from X in the following strong sense: $\mathbf{x}_0$ belongs to the open half-space H_{11} and X is contained in the open half-space H_{22}. This is illustrated in Figure 10.3.

The reason for the 'finite' in the Finite Separation Theorem is of course, that X is the convex hull of a finite set. But notice that if we enlarge X in Figure 10.3 to include all points on and within the dotted curve, strong separation still obtains. Results of this kind have been formalised into *separation theorems*, which play an important

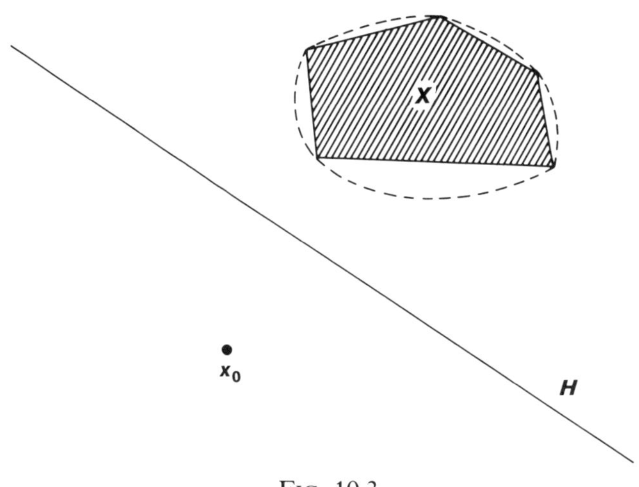

FIG. 10.3

rôle in mathematical economics. We do not discuss them here: there are several subtly different separation theorems, and to know which is applicable when requires a rather deeper appreciation of the properties of closed and bounded sets than we have attempted to convey in this book.

Exercises

1. Which of the following sets are convex?

$$\{\mathbf{x} \in R^2 : x_2 > x_1^2\}$$
$$\{\mathbf{x} \in R^2 : |x_2| > x_1^2\}$$
$$\{\mathbf{x} \in R^3 : |x_1 + x_2 + x_3| > 1\}$$
$$\{\mathbf{x} \in R^4 : x_1 > x_2 \text{ and } x_3^2 + x_4^2 < 1\}$$

2. A set K in R^n is said to be a *cone* if $\alpha\mathbf{x} \in K$ whenever $\mathbf{x} \in K$ and $\alpha > 0$. Prove the following two propositions.
   (i) A cone K is convex if and only if $\mathbf{x} + \mathbf{y} \in K$ for all $\mathbf{x}$ and $\mathbf{y}$ in K.
   (ii) A set X is a convex cone if and only if $\lambda\mathbf{x} + \mu\mathbf{y} \in X$ whenever $\mathbf{x}$ and $\mathbf{y}$ are in X and λ and μ are positive.

   Give two examples of cones in R^2, one convex and the other not.

3. Show that the convex hull of a sphere is a closed ball.

10.2 Concavity

Let f be a function on a set X in R^n. We define the *hypograph* of f to be the set

$$\text{hyp } f = \{(\mathbf{x}: \lambda) \in R^{n+1}: \mathbf{x} \in X \quad \text{and} \quad \lambda \leq f(\mathbf{x})\}.$$

Thus if $n = 1$, hyp f is the set of all points in the plane which lie on or below the graph of f ('hypo' is Greek for below).

A function f defined on a set X in R^n is said to be a *concave function* if its hypograph is a convex set in R^{n+1}: notice that this definition *implies* that the domain X of f be convex, so concave functions live on convex sets. In particular, a function of one real variable is concave if and only if the set of points in the plane which lie on or below its graph is a convex set.

The algebraic description of a concave function runs as follows.

THEOREM 1 A function f defined on a convex set X in R^n is concave if and only if

$$f(\alpha\mathbf{x} + (1 - \alpha)\mathbf{y}) \geq \alpha f(\mathbf{x}) + (1 - \alpha)f(\mathbf{y})$$

whenever $\mathbf{x}$ and $\mathbf{y}$ are in X and $0 < \alpha < 1$.

PROOF The hypograph of a concave function f is a convex set which owns all points of the form $(\mathbf{x}: f(\mathbf{x}))$; the 'only if' part of the theorem follows immediately. To prove 'if', suppose f is not concave. Then we may choose two points in hyp f such that the line segment joining them is not entirely contained in hyp f. This means that there exist members $\mathbf{x}$ and $\mathbf{y}$ of X, and scalars α, λ, μ, such that $0 < \alpha < 1$, $f(\mathbf{x}) \geq \lambda$, $f(\mathbf{y}) \geq \mu$ and

$$f(\alpha\mathbf{x} + (1 - \alpha)\mathbf{y}) < \alpha\lambda + (1 - \alpha)\mu.$$

But then $f(\alpha\mathbf{x} + (1 - \alpha)\mathbf{y}) < \alpha f(\mathbf{x}) + (1 - \alpha)f(\mathbf{y})$. QED

A function f on R^n is said to be *affine* if

$$f(\mathbf{x}) = \mathbf{p}^{\mathrm{T}}\mathbf{x} - \beta$$

for some given n-vector $\mathbf{p}$ and scalar β. Notice that this includes the cases $\mathbf{p} = \mathbf{0}$ and $\beta = 0$: thus constant functions are affine, and so are linear forms. If f is affine, the weak inequality of Theorem 1 becomes an equation: thus affine functions are a very special case of concave functions. We shall have more to say about this at the end of the section. To obtain less special examples of concave functions, we introduce the following theorem.

THEOREM 2 Let f and g be concave functions on a convex set X in R^n, and let

$$h(\mathbf{x}) = \min\ (f(\mathbf{x}), g(\mathbf{x})).$$

Then h is a concave function on X.

PROOF The hypograph of h is the intersection of the hypographs of f and g, so h is concave by Theorem 10.1.1. QED

To see how Theorem 2 may be applied, define a function h on R^3 by setting

$$h(\mathbf{x}) = \min\ (x_1 + x_2 + 2, x_2 + x_3 - 3).$$

Using Theorem 2 and the fact that affine functions are concave, we see that h is concave. More generally, let $f_1, \ldots, f_k$ be affine functions on R^n and let

$$f(\mathbf{x}) = \min\ (f_1(\mathbf{x}), f_2(\mathbf{x}), \ldots, f_k(\mathbf{x})).$$

Then f is concave. It may be shown that the valuation function $v(\mathbf{b})$ in linear programming is always of this type: a rigorous proof of this requires a deeper consideration of duality and the simplex method than has been given in this book.

We shall give more examples of concave functions on R^n before very long, but it is worthwhile first to devote some attention to the case $n = 1$. Since concave functions live on convex sets, concave functions of one variable live on intervals. The basic theorem about such functions is Theorem 3, which is illustrated in Figures 10.4 and 10.5.

THEOREM 3 Let f be a function defined on an interval I in R. Then f is concave if and only if

$$\frac{f(x_2) - f(x_1)}{x_2 - x_1} \geq \frac{f(x_3) - f(x_2)}{x_3 - x_2} \tag{1}$$

for any x_1, x_2, x_3 in I such that $x_1 < x_2 < x_3$.

PROOF Given that $x_1 < x_2 < x_3$, (1) states that

$$(x_3 - x_1)f(x_2) \geq (x_3 - x_2)f(x_1) + (x_2 - x_1)f(x_3).$$

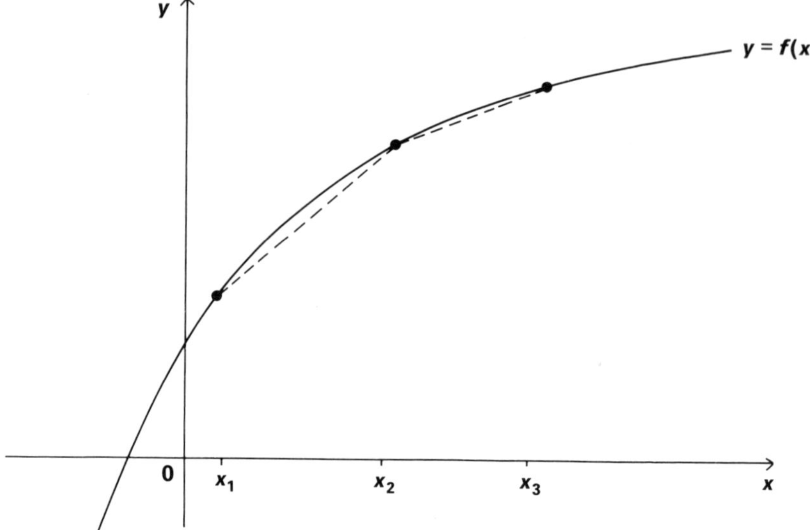

FIG. 10.4 f is concave

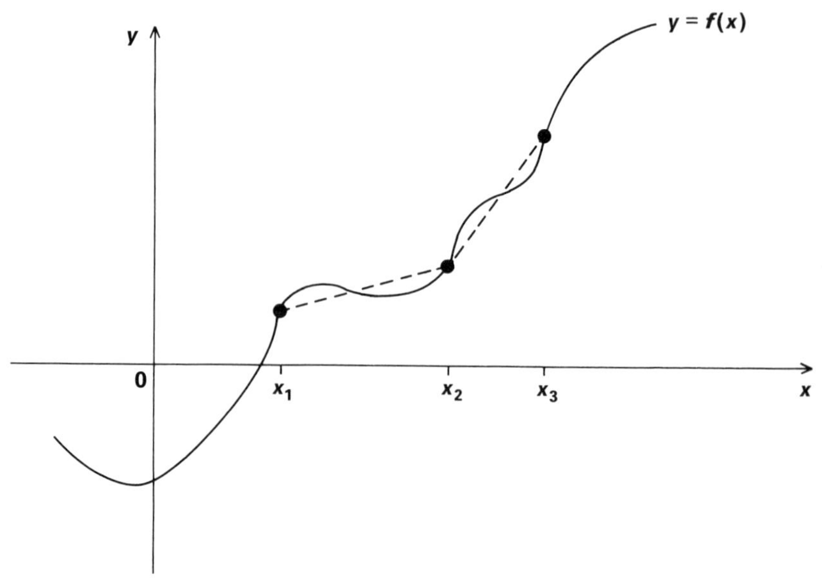

FIG. 10.5 f is not concave

This in turn is equivalent to the statement that

$$f(x_2) \geq \alpha f(x_1) + (1 - \alpha) f(x_3) \tag{2}$$

where $\alpha = (x_3 - x_2)/(x_3 - x_1)$. With this definition of α,

$$0 < \alpha < 1 \quad \text{and} \quad x_2 = \alpha x_1 + (1 - \alpha) x_3. \tag{3}$$

If f is concave and $x_1 < x_2 < x_3$, we can choose a number α satisfying (3) and infer (2) from Theorem 1: hence (1) holds. Conversely, suppose that f satisfies (1) whenever $x_1 < x_2 < x_3$. Let $x_1 < x_3$ and let $0 < \alpha < 1$: then we can use (3) to define a number x_2, and infer (2) from (1). Hence (2) holds whenever $x_1 < x_3$ and $0 < \alpha < 1$, so f is concave by Theorem 1. QED

We now give some conditions for concavity of functions of one variable in terms of first derivatives $f'(x)$ and second derivatives $f''(x)$.

THEOREM 4 Let f be a differentiable function defined on an interval I in R. Then f is concave if and only if

$$f'(u) \geq f'(v)$$

for any u and v in I such that $u < v$.

PROOF Let f be concave and let $u < v$. Let δ be a positive number such that $2\delta < v - u$. Applying Theorem 3 twice,

$$\frac{f(u + \delta) - f(u)}{\delta} \geq \frac{f(v - \delta) - f(u + \delta)}{v - u - 2\delta} \geq \frac{f(v) - f(v - \delta)}{\delta}.$$

Letting δ approach zero we have $f'(u) \geq f'(v)$.

Conversely, suppose that f is not concave. Then we may choose x_1, x_2, x_3 in I such that $x_1 < x_2 < x_3$ and (1) is false. By the Mean Value Theorem there exist numbers u and v such that

$$\begin{aligned}
&x_1 < u < x_2 < v < x_3,\\
&(x_2 - x_1) f'(u) = f(x_2) - f(x_1),\\
&(x_3 - x_2) f'(v) = f(x_3) - f(x_2).
\end{aligned}$$

Then $u < v$ and $f'(u) < f'(v)$. QED

THEOREM 5 A twice-differentiable function f on an interval I in R is concave if and only if $f''(x) \leq 0$ for all x in I.

PROOF Immediate from Theorem 4. QED

Example 1 Let θ be a number such that $\theta \leq 1$ and $\theta \neq 0$. For all positive x, let

$$f(x) = (x^\theta - 1)/\theta.$$

Then $f'(x) = x^{\theta-1}$ so $f''(x) = (\theta - 1)x^{\theta-2} \leq 0$. By Theorem 5, f is concave.

Example 2 Let $f(x) = \ln x$ for all positive x. Here 'ln' denotes natural logarithm or '$\log_e$'. Recall from elementary calculus that ln is the inverse function of exp, where

$$\exp t = e^t = 1 + t + t^2/2! + t^3/3! + \cdots.$$

In this case, $f'(x) = x^{-1}$ and $f''(x) = -x^{-2}$, so f is concave by Theorem 5.

In fact, this f is the limit of the f of Example 1 as θ approaches zero. To see this, fix a positive number x and let $\ln x = t$. Then $x^\theta = \exp(\theta t)$ so

$$(x^\theta - 1)/\theta = t + \theta t^2/2! + \theta^2 t^3/3! + \cdots$$

which approaches t as θ approaches zero.

We now return to concave functions of n variables. The n-dimensional analogue of Theorem 5 is as follows.

THEOREM 6 Let f be a C^2 function defined on an open convex set U in R^n. For each $\mathbf{x}$ in U define the *Hessian matrix* of f at $\mathbf{x}$ to be the $n \times n$ matrix $\mathbf{H}(\mathbf{x}) = (h_{ij}(\mathbf{x}))$ such that

$$h_{ij}(\mathbf{x}) = \partial^2 f(\mathbf{x})/\partial x_i \, \partial x_j \quad \text{for all } i, j.$$

Then f is concave if and only if $\mathbf{a}^T\mathbf{H}(\mathbf{x})\mathbf{a} \leq 0$ for every $\mathbf{x}$ in U and every $\mathbf{a}$ in R^n.

Theorem 6, which will not be proved here, provides a test for whether or not a C^2 function is concave. The test is often hard to apply but one way of implementing it, using determinants, is discussed at the end of Appendix C. Notice that Theorem 6 is an 'if and only if' result and can therefore be used to show that a given function is *not* concave.

The easy way of constructing functions of n variables which *are* concave is to build them up from concave functions of one variable. Notice that if $f(\mathbf{x})$ is a function on R^n which happens to depend only on x_1, and is a concave function of x_1, then $f(\mathbf{x})$ is a concave function; this is easily proved using either the original definition of concavity or Theorem 1.

There are four main 'building-up' theorems, of which the first is Theorem 2. The other three are as follows.

THEOREM 7 Let f be a concave function on a convex set X in R^n and let $g(\mathbf{x}) = \lambda f(\mathbf{x})$, where λ is a positive scalar. Then g is concave.

THEOREM 8 Let f and g be concave functions on a convex set X in R^n and let $h(\mathbf{x}) = f(\mathbf{x}) + g(\mathbf{x})$. Then h is concave.

THEOREM 9 Let f be a concave function on a convex set X in R^n. Let I be an interval in R such that $f(\mathbf{x}) \in I$ for all $\mathbf{x}$ in X. Let ϕ be a *non-decreasing* concave function on I and let $g(\mathbf{x}) = \phi(f(\mathbf{x}))$ for all $\mathbf{x}$ in X.

Then g is a concave function.

The proofs of Theorems 7 and 8 are simple applications of Theorem 1 and are left as exercises for the reader. To prove Theorem 9 let X, I, f, ϕ and g be as given: also let $\mathbf{x}$ and $\mathbf{y}$ be members of X, and let $0 < \alpha < 1$. Since f is concave and ϕ is non-decreasing,

$$\phi(f(\alpha\mathbf{x} + (1 - \alpha)\mathbf{y})) \geq \phi(\alpha f(\mathbf{x}) + (1 - \alpha)f(\mathbf{y})).$$

But ϕ is concave, so

$$\phi(\alpha f(\mathbf{x}) + (1 - \alpha)f(\mathbf{y})) \geq \alpha\phi(f(\mathbf{x})) + (1 - \alpha)\phi(f(\mathbf{y})).$$

Putting these inequalities together we have

$$g(\alpha\mathbf{x} + (1 - \alpha)\mathbf{y}) \geq \alpha g(\mathbf{x}) + (1 - \alpha)g(\mathbf{y}).$$

Example 3 Let $\theta_1, \ldots, \theta_n, \lambda_1, \ldots, \lambda_n$ be numbers such that for $i = 1, \ldots, n$,

$$\theta_i \leq 1, \quad \theta_i \neq 0 \quad \text{and } \lambda_i \text{ has thc samc sign as } \theta_i.$$

Define a function f on the positive orthant P^n by setting

$$f(\mathbf{x}) = \lambda_1 x_1^{\theta_1} + \lambda_2 x_2^{\theta_2} + \cdots + \lambda_n x_n^{\theta_n}.$$

(For example, with $n = 3$ we could have $f(\mathbf{x}) = 2x_1 - 3/x_2 + 4\sqrt{x_3}$.)

We show that f is concave. For $i = 1, \ldots, n$, let

$$f_i(\mathbf{x}) = (x_i^{\theta_i} - 1)/\theta_i, \quad \beta_i = \lambda_i\theta_i.$$

Since λ_i and θ_i have the same sign, $\beta_i > 0$. Also

$$f(\mathbf{x}) = \beta_0 + \beta_1 f_1(\mathbf{x}) + \cdots + \beta_n f_n(\mathbf{x})$$

where $\beta_0 = \lambda_1 + \cdots + \lambda_n$. By Example 1 and Theorems 7 and 8, f is concave.

Example 4 Let f be as in Example 3 and suppose in addition that $\theta_i > 0$ for all i. Then $f(\mathbf{x}) > 0$ for all $\mathbf{x}$ in P^n so we may define a function g on P^n by setting $g(\mathbf{x}) = \ln f(\mathbf{x})$. By Theorem 9 and Examples 2 and 3, g is concave.

Example 5 Define the function f on P^3 by

$$f(\mathbf{x}) = \min\,(x_2 + \ln x_3,\, x_3 + \ln x_1,\, x_1 + \ln x_2).$$

By Example 2 and Theorems 2 and 8, f is concave.

A function f defined on a convex set X in R^n is said to be *convex* if $-f$ is concave. Thus a convex function is one that has a convex *epigraph*, where 'epi' is Greek for 'above'. Translating statements about concave functions into statements about convex functions is a routine exercise which does require a modicum of care. For example, Theorem 9 remains true when 'concave' is replaced by 'convex' each time it appears (three times): notice that 'non-decreasing' is unchanged in translation. On the other hand, the convex-function version of Theorem 2 replaces 'concave' by 'convex' (twice) *and* 'min' by 'max'. To carry on in this vein would be boring, and we do not want to question the reader's ability to multiply by -1. But, to repeat, care is required.

We end this section with a theorem about affine functions which will be used in the next chapter.

THEOREM 10 A function on R^n is affine if and only if it is both concave and convex.

PROOF 'Only if' is obvious. To prove 'if' let f be both concave and convex and define the function $g(\mathbf{x}) = f(\mathbf{x}) - f(\mathbf{0})$. Then

$$g(\mathbf{0}) = 0 \tag{4}$$

and g is both concave and convex. Thus

$$g(\alpha\mathbf{x} + (1-\alpha)\mathbf{y}) = \alpha g(\mathbf{x}) + (1-\alpha)g(\mathbf{y}) \tag{5}$$

whenever $\mathbf{x}$ and $\mathbf{y}$ are in R^n and $0 < \alpha < 1$. Letting $\mathbf{y} = \mathbf{0}$ in (5) and applying (4), we have $g(\alpha\mathbf{x}) = \alpha g(\mathbf{x})$, which may also be written $g(\mathbf{x}) = \alpha^{-1} g(\alpha\mathbf{x})$. Thus

$$g(\lambda\mathbf{x}) = \lambda g(\mathbf{x}) \tag{6}$$

for all positive λ and, by (4), for $\lambda = 0$ also.

Now apply (5) with $\alpha = \frac{1}{2}$, multiply through by 2 and apply (6). We obtain

$$g(\mathbf{x} + \mathbf{y}) = g(\mathbf{x}) + g(\mathbf{y}). \tag{7}$$

Setting $\mathbf{y} = -\mathbf{x}$ in (7) and applying (4), we have $g(-\mathbf{x}) = -g(\mathbf{x})$. Thus (6) holds for all real λ. This, together with (7), implies that

$$g(\lambda\mathbf{x} + \mu\mathbf{y}) = \lambda g(\mathbf{x}) + \mu g(\mathbf{y}) \tag{8}$$

for all $\mathbf{x}$ and $\mathbf{y}$ and all real λ and μ.

For $j = 1, \ldots, n$, let $\mathbf{u}_j$ denote column j of $\mathbf{I}_n$ and let $p_j = g(\mathbf{u}_j)$. By (8), $g(\mathbf{x}) = \mathbf{p}^{\mathrm{T}}\mathbf{x}$ for all $\mathbf{x}$. Thus $f(\mathbf{x})$ is the affine function

$$\mathbf{p}^{\mathrm{T}}\mathbf{x} + f(\mathbf{0}).$$ QED

Exercises

1. Using only one diagram, sketch the functions of Examples 1 and 2 for several different values of θ.

2. Let A and B be sets in R^n with A convex; B need not be convex. Suppose that for each member $\mathbf{p}$ of A, the problem

   minimise $\mathbf{p}^{\mathrm{T}}\mathbf{x}$ subject to $\mathbf{x} \in B$

   is soluble: denote the solution value by $f(\mathbf{p})$.

   Show that f is a concave function on A.

   This result has been interpreted to imply that 'consumers like stable prices while producers like fluctuating prices'. Why?

3. Let f, g, h be concave functions on R^n. Let $\hat{\mathbf{x}}$ be a solution vector for the problem

   maximise $f(\mathbf{x})$ subject to $g(\mathbf{x}) \geq 0$

   and let $\mathbf{x}^*$ be a solution vector for the problem

   maximise $f(\mathbf{x})$ subject to $g(\mathbf{x}) \geq 0$ and $h(\mathbf{x}) \geq 0$.

   Show that if $h(\hat{\mathbf{x}}) < 0$ then $h(\mathbf{x}^*) = 0$.

   (This does not require any of the results of the next chapter.)

4. Show that a differentiable function f on R is concave if and only if

$$f(x+a) \leq f(x) + af'(x) \quad \text{for all } x \text{ and } a.$$

5. A function f on a convex set X in R^n is said to be *strictly concave* if

$$f(\alpha\mathbf{x} + (1-\alpha)\mathbf{y}) > \alpha f(\mathbf{x}) + (1-\alpha)f(\mathbf{y})$$

whenever $\mathbf{x} \in X$, $\mathbf{y} \in X$, $\mathbf{x} \neq \mathbf{y}$ and $0 < \alpha < 1$.
(i) Let f be a differentiable function on R. Show that f is strictly concave if and only if f' is a decreasing function.
(ii) Give an example of a twice-differentiable, strictly concave function f on R such that $f''(x) = 0$ for infinitely many values of x.
(iii) On P^3 define $f(\mathbf{x}) = \sqrt{x_1} + 2\sqrt{x_2} + 3\sqrt{x_3}$. Show that f is strictly concave.

10.3 Quasi-concavity

As with concave functions, we give a geometrical definition and an algebraic one. A function f defined on a convex set X in R^n is said to be *quasi-concave* if, for any number β, the set

$$\{\mathbf{x} \in X : f(\mathbf{x}) \geq \beta\}$$

is a convex set in R^n.

THEOREM 1 A function f on a convex set X in R^n is quasi-concave if and only if

$$f(\alpha\mathbf{x} + (1-\alpha)\mathbf{y}) \geq \min\,(f(\mathbf{x}), f(\mathbf{y}))$$

whenever $\mathbf{x}$ and $\mathbf{y}$ are in X and $0 < \alpha < 1$.
PROOF Suppose that f is quasi-concave, let $\mathbf{x}$ and $\mathbf{y}$ be members of X and let

$$\beta = \min\,(f(\mathbf{x}), f(\mathbf{y})).$$

Then the set $\{\mathbf{z} \in X : f(\mathbf{z}) \geq \beta\}$ is convex and owns $\mathbf{x}$ and $\mathbf{y}$, which proves 'only if'.

Conversely, suppose that f is not quasi-concave. Then we can choose members $\mathbf{x}$ and $\mathbf{y}$ of X, and numbers α and β, such that

$$f(\mathbf{x}) \geq \beta, \quad f(\mathbf{y}) \geq \beta, \quad 0 < \alpha < 1$$

and $f(\alpha\mathbf{x} + (1-\alpha)\mathbf{y}) < \beta$. But then

$$\min\,(f(\mathbf{x}), f(\mathbf{y})) > \beta > f(\alpha\mathbf{x} + (1-\alpha)\mathbf{y}). \qquad \text{QED}$$

If f is a concave function on a convex set X then the set

$$\{\mathbf{x} \in X: (\mathbf{x}: \beta) \in \text{hyp } f\}$$

is convex for every β. Thus any concave function is quasi-concave. On the other hand, there exist quasi-concave functions which are not concave. To see this, notice that Theorem 1 implies that *any non-decreasing function of one real variable is quasi-concave.* Thus $f(x) = x^3$ is a quasi-concave function on R which is not concave.

The three main theorems which enable us to construct quasi-concave functions are as follows.

THEOREM 2 Let f and g be quasi-concave functions on a convex set X in R^n and let

$$h(\mathbf{x}) = \min (f(\mathbf{x}), g(\mathbf{x})).$$

Then h is a quasi-concave function on X.

THEOREM 3 Let f be a quasi-concave function on a convex set X in R^n and let $g(\mathbf{x}) = \lambda f(\mathbf{x})$, where λ is a positive scalar. Then g is quasi-concave.

THEOREM 4 Let f be a quasi-concave function on a convex set X in R^n. Let I be an interval in R such that $f(\mathbf{x}) \in I$ for all $\mathbf{x}$ in X. Let ϕ be a non-decreasing function on I and let $g(\mathbf{x}) = \phi(f(\mathbf{x}))$ for all $\mathbf{x}$ in X.

Then g is a quasi-concave function.

Theorem 2 follows immediately from the fact that the intersection of two convex sets is convex. Theorem 3 is obvious. To prove Theorem 4, let $\mathbf{x}$ and $\mathbf{y}$ be members of X, let $0 < \alpha < 1$ and let $\mathbf{z} = \alpha\mathbf{x} + (1 - \alpha)\mathbf{y}$. By Theorem 1, we have $f(\mathbf{z}) \geq \lambda$, where

$$\lambda = \min (f(\mathbf{x}), f(\mathbf{y})).$$

Since ϕ is non-decreasing,

$$\phi(f(\mathbf{z})) \geq \phi(\lambda) = \min (\phi(f(\mathbf{x})), \phi(f(\mathbf{y}))).$$

Thus $g(\mathbf{z}) \geq \min (g(\mathbf{x}), g(\mathbf{y}))$ and g is quasi-concave.

Notice that Theorems 2 and 3 are just Theorems 2 and 7 of the last section with 'concave' replaced by 'quasi-concave' throughout. Similarly, since any non-decreasing function of one variable is quasi-concave, Theorem 4 is just Theorem 10.2.9 with 'concave' replaced by 'quasi-concave' throughout. No such trick may be played with

Theorem 10.2.8: if f and g are quasi-concave then $f+g$ may not be quasi-concave. To see this, notice that Theorem 1 implies that non-increasing functions of one variable, as well as non-decreasing ones, are quasi-concave. Thus

$$f(x) = \max(x, 0) \quad \text{and} \quad g(x) = \max(-x, 0)$$

are quasi-concave functions. But $f(x)+g(x) = |x|$, which is not a quasi-concave function:

$$|2| = |-2| > 1 > |0|.$$

We now give two examples of quasi-concave functions on the non-negative orthant R^2_+.

Example 1 Define f on R^2_+ by $f(\mathbf{x}) = (x_1 + 2) \exp x_2$.

Then $f(\mathbf{x}) = \exp g(\mathbf{x})$ where $g(\mathbf{x}) = \ln(x_1 + 2) + x_2$. Using the methods of the last section, it is easy to show that g is concave. Since the exponential function is increasing, f is quasi-concave by Theorem 4.

Example 2 Define h on R^2_+ by $h(\mathbf{x}) = \min(3x_1 - 2x_2, x_2^2)$.

Then we may write $h = \min(f, g)$ where f is affine and $g(\mathbf{x})$ is an increasing function of x_2 alone. So by Theorem 2, h is quasi-concave.

In economics, production functions and utility functions are conventionally assumed to be quasi-concave. Production functions are often assumed also to be homogeneous, particularly in applied work: the degree of homogeneity is then a parameter measuring returns to scale. The last theorem of this section demonstrates a useful relationship between homogeneity, quasi-concavity and concavity. In Examples 3 and 4, we apply this to two functions much used in economics.

Let f be a function on the positive orthant P^n and let τ be a positive number. We say that f is *positive-homogeneous* of degree τ if

$$f(\lambda\mathbf{x}) = \lambda^\tau f(\mathbf{x}) > 0$$

for all $\mathbf{x}$ in P^n and every positive λ. Theorem 5 states, in effect, that a homogeneous quasi-concave production function is concave if and only if it exhibits constant or diminishing returns to scale.

THEOREM 5 Let the function f on P^n be quasi-concave and positive-homogeneous of degree τ. Then f is concave if and only if $\tau \leq 1$.

PROOF First let $\tau > 1$, let $\mathbf{e}$ be the n-vector consisting entirely of 1's, and let $f(\mathbf{e}) = \mu$. Then $\mu > 0$, and $f(\lambda\mathbf{e}) = \lambda^\tau \mu$ for every positive λ. Since $\tau > 1$, $1 + 3^\tau > 2^{\tau+1}$ so $f(\mathbf{e}) + f(3\mathbf{e}) > 2f(2\mathbf{e})$. Thus f is not concave.

Now let $\tau \leq 1$. Then we can write $f(\mathbf{x}) = (g(\mathbf{x}))^\tau$ where $g(\mathbf{x}) = (f(\mathbf{x}))^{1/\tau}$. If we can show that g is concave then f is also concave by Theorem 10.2.9. But g is positive-homogeneous of degree 1. Hence we may as well assume that $\tau = 1$.

Assume this and let $\mathbf{x}$ and $\mathbf{y}$ be members of P^n. Let $0 < \alpha < 1$ and set $\mathbf{z} = \alpha\mathbf{x} + (1 - \alpha)\mathbf{y}$. Define the positive numbers

$$\lambda = f(\mathbf{x}), \quad \mu = f(\mathbf{y}), \quad v = \alpha\lambda + (1 - \alpha)\mu.$$

We wish to show that $f(\mathbf{z}) \geq v$.

Now $\mathbf{z} = \alpha\lambda(\lambda^{-1}\mathbf{x}) + (1 - \alpha)\mu(\mu^{-1}\mathbf{y})$. Multiplying by v^{-1} we have

$$v^{-1}\mathbf{z} = \theta\lambda^{-1}\mathbf{x} + (1 - \theta)\mu^{-1}\mathbf{y} \quad \text{where } \theta = \alpha\lambda/v.$$

Since $\tau = 1$, $f(\lambda^{-1}\mathbf{x}) = f(\mu^{-1}\mathbf{y}) = 1$. Also, $0 < \theta < 1$, so $f(v^{-1}\mathbf{z}) \geq 1$ by quasi-concavity. Appealing again to the assumption that $\tau = 1$, we have the required result. QED

Example 3 (Cobb–Douglas Production Function)
Let A and $\alpha_1, \ldots, \alpha_n$ be positive numbers. Define the function f on P^n by setting

$$f(\mathbf{x}) = Ax_1^{\alpha_1}x_2^{\alpha_2} \cdots x_n^{\alpha_n}.$$

Then $f(\mathbf{x}) = A \exp g(\mathbf{x})$ where

$$g(\mathbf{x}) = \alpha_1 \ln x_1 + \alpha_2 \ln x_2 + \cdots \alpha_n \ln x_n.$$

Clearly, g is concave, so f is quasi-concave. f is homogeneous of degree

$$\tau = \alpha_1 + \alpha_2 + \cdots + \alpha_n$$

so f is concave if and only if $\tau \leq 1$.

Example 4 (CES Production Function)
Let A, τ and $\delta_1, \ldots, \delta_n$ be positive numbers such that

$$\delta_1 + \cdots + \delta_n = 1.$$

Let θ be a number such that $\theta \leq 1$ and $\theta \neq 0$. Define the function f on P by setting

$$f(\mathbf{x}) = A(\delta_1 x_1^\theta + \delta_2 x_2^\theta + \cdots + \delta_n x_n^\theta)^{\tau/\theta}.$$

We now show that f is quasi-concave. Define the function g on P by setting

$$g(\mathbf{x}) = (\delta_1 x_1^\theta + \delta_2 x_2^\theta + \cdots \delta_n x_n^\theta)/\theta.$$

Since the δ's are positive and $\theta \leq 1$, g is a sum of concave functions and therefore concave. Define the interval I to be the set of all positive numbers if $\theta > 0$, the set of all negative numbers if $\theta < 0$. In each case, $g(\mathbf{x}) \in I$ for each $\mathbf{x}$ in P and the function

$$\phi(u) = (u/\theta)^{\tau/\theta}$$

is increasing on I. Also $f(\mathbf{x}) = \phi(g(\mathbf{x}))$ for all $\mathbf{x}$ in P, so f is quasi-concave by Theorem 4.

Notice that f is positive-homogeneous of degree τ, so f is concave if and only if $\tau \leq 1$.

We now show that the limit of this function f as θ approaches zero is the function f of Example 3, with $\alpha_i = \delta_i \tau$ for all i. To prove this, recall from the last section that

$$(y^\theta - 1)/\theta \to \ln y \quad \text{as} \quad \theta \to 0 \tag{1}$$

for any positive y. Inverting this result, we have

$$(1 + \theta z)^{1/\theta} \to \exp z \quad \text{as} \quad \theta \to 0 \tag{2}$$

for any real z; indeed, the exponential function is defined in some calculus textbooks as this limit. Since the δ's sum to 1, the function f which concerns us here may be written

$$f(\mathbf{x}, \theta) = A\left(1 + \theta \sum_{i=1}^{n} (\delta_i/\theta)(x_i^\theta - 1)\right)^{\tau/\theta}.$$

Letting $\theta \to 0$ and applying (1) and (2) we see that

$$\begin{aligned}\lim_{\theta \to 0} f(\mathbf{x}, \theta) &= A \lim_{\theta \to 0}\left(1 + \theta \sum_{i=1}^{n} \delta_i \ln x_i\right)^{\tau/\theta} \\ &= A\left(\exp\left(\sum_{i=1}^{n} \delta_i \ln x_i\right)\right)^{\tau} \\ &= A(x_1^{\delta_1} x_2^{\delta_2} \cdots x_n^{\delta_n})^{\tau}\end{aligned}$$

The initials CES stand for constant elasticity of substitution: the elasticity of substitution of $f(\mathbf{x}, \theta)$ is $1/(1 - \theta)$, so the limiting result says that the CES function approaches the Cobb–Douglas function as the elasticity of substitution approaches 1. For the economics of elasticities of substitution see Dixit (1976) pp. 78–80.

Our final remarks in this section are similar to those at the end of the last one. We define a function f on a convex set X in R^n to be *quasi-convex* if $-f$ is quasi-concave. Theorems 1–4 translate easily into statements about quasi-convex functions.

Translating Theorem 5 is a little harder: for if f takes only positive values then $-f$ cannot be positive-homogeneous of any degree. One correct translation of Theorem 5 runs as follows: a function which is quasi-convex and positive-homogeneous of degree τ on P^n is convex if and only if $\tau \geq 1$.

Exercises

1. Given the function f of Example 1, sketch the curves $f(x_1, x_2) = \beta$ for $\beta = 4, 6, 10$.
2. Given the function h of Example 2, sketch the curves $h(x_1, x_2) = \beta$ for $\beta = 1, 4, 9$.
3. For positive x_1 and x_2, let
$$f(x_1, x_2) = (3x_1^{1/4} + 4x_1^{1/2})x_2^{1/2}.$$
Show that f is concave.
4. Let α be positive and let $f(\mathbf{x}) = \|\mathbf{x}\|^\alpha$ for all $\mathbf{x}$ in R^n. Show that f is quasi-convex, being convex if $\alpha \geq 1$.
5. A function f on a convex set X in R^n is said to be *strictly quasiconcave* if
$$f(\alpha\mathbf{x} + (1 - \alpha)\mathbf{y}) > \min\,(f(\mathbf{x}), f(\mathbf{y}))$$
whenever $\mathbf{x} \in X$, $\mathbf{y} \in X$, $\mathbf{x} \neq \mathbf{y}$ and $0 < \alpha < 1$.
(i) Show that a strictly concave function is strictly quasi-concave.
(ii) Show that a strictly quasi-concave function has at most one maximum point.
(iii) Show that a strictly quasi-concave function which is positive-homogeneous of degree τ on P^n is strictly concave if and only if $\tau < 1$.

11 Concave Programming

In Section 11.1 we bring together the analysis of the last two chapters to derive necessary and sufficient conditions for constrained maxima when all relevant functions are C^1 and concave. In the following section we discuss the extent to which the theorems remain valid when the C^1 assumption is dropped. We end the chapter with some remarks about duality in concave programming.

The method of Section 11.1 is differential calculus and we shall be fairly rigorous, appealing to the theorems of Chapter 9. We have not provided proofs of the main results in Section 11.2: such would require a thorough discussion of the separation theorems mentioned at the end of Section 10.1.

11.1 Conditions for a maximum

The reason why concave functions are important in the theory of maximisation is that the usual necessary condition for a local maximum becomes a necessary and sufficient condition for a maximum when the function in question is concave.

THEOREM 1 Let f be a concave C^1 function defined on an open convex set V in R^n. Then f has a maximum at $\hat{\mathbf{x}}$ if and only if $\nabla f(\hat{\mathbf{x}}) = \mathbf{0}$.

PROOF Any maximum is a local maximum, so the 'only if' part is implied by Theorem 9.2.1 and the remarks following it. To prove 'if', let $\mathbf{x}$ and $\mathbf{y}$ be points of V such that $f(\mathbf{x}) > f(\mathbf{y})$ and let $\mathbf{z} = \nabla f(\mathbf{y})$. We want to show that $\mathbf{z} \neq \mathbf{0}$.

Since f is concave,

$$f(\alpha\mathbf{x} + (1 - \alpha)\mathbf{y}) \geq \alpha f(\mathbf{x}) + (1 - \alpha)f(\mathbf{y})$$

whenever $0 < \alpha < 1$. Let $\mathbf{q} = \mathbf{x} - \mathbf{y}$, $\beta = f(\mathbf{x}) - f(\mathbf{y})$. Then

$$f(\mathbf{y} + \alpha\mathbf{q}) - f(\mathbf{y}) \geq \alpha\beta$$

for all sufficiently small positive α, so $\mathbf{q}^{\mathrm{T}}\mathbf{z} \geq \beta > 0$. Thus $\mathbf{z} \neq \mathbf{0}$.

QED

We now turn to the main topic of this chapter. There are many different formulations for concave programming, and we shall use the following one:

(P) maximise $f_0(\mathbf{x})$ subject to $f_i(\mathbf{x}) \geq 0$ for $i = 1, \ldots, m$.

Here $f_0, f_1, \ldots, f_m$ are concave functions defined on a convex set X in R^n. The requirement that $\mathbf{x} \in X$ is called the *implicit constraint* of (P). The terms solution vector and solution value will be used in the same sense as in linear programming.

In view of Theorem 1, we can strengthen the statements of the Fritz John and Kuhn–Tucker Theorems in the concave case, replacing statements about gradients being zero by statements about functions being maximised.

To explain what is going on, and to avoid the ambiguous term 'regular maximum', we make the following non-standard definitions. We say that the n-vector $\hat{\mathbf{x}}$ in X is a *Fritz John point* for the problem (P) if there exist multipliers $\phi_0, \phi_1, \ldots, \phi_m$ with the following properties:

(I) $\sum_{i=0}^{m} \phi_i f_i(\hat{\mathbf{x}}) \geq \sum_{i=0}^{m} \phi_i f_i(\mathbf{x})$ for all $\mathbf{x}$ in X;

(II) for each $i \geq 1$, the numbers ϕ_i and $f_i(\hat{\mathbf{x}})$ are non-negative and at least one of them is zero;

(III) $\phi_0 \geq 0$ and $\phi_0, \phi_1, \ldots, \phi_m$ are not all zero.

We say that the Fritz John point $\hat{\mathbf{x}}$ is a *Kuhn–Tucker point* if the multipliers $\phi_0, \phi_1, \ldots, \phi_m$ can be chosen such that $\phi_0 > 0$. In this case we can set

$$\lambda_1 = \phi_1/\phi_0, \ldots, \lambda_m = \phi_m/\phi_0,$$

inferring that (I) and (II) hold with ϕ_i replaced by λ_i for $i \geq 1$ and ϕ_0 by 1.

Thus the point $\hat{\mathbf{x}}$ of X is a Kuhn–Tucker point for (P) if and only if there exist multipliers $\lambda_1, \ldots, \lambda_m$ such that

(I*) the function $f_0(\mathbf{x}) + \sum_{i=1}^{m} \lambda_i f_i(\mathbf{x})$ has a maximum at $\hat{\mathbf{x}}$;

(II*) the m-vectors

$$\begin{pmatrix} \lambda_1 \\ \vdots \\ \lambda_m \end{pmatrix} \text{ and } \begin{pmatrix} f_1(\hat{\mathbf{x}}) \\ \vdots \\ f_m(\hat{\mathbf{x}}) \end{pmatrix}$$

are non-negative and complementary.

We now state and prove one small theorem and one big one on how Fritz John points, Kuhn–Tucker points and solution vectors are related. The 'sufficiency theorem' (Theorem 2) is school algebra. Theorem 3 is the main theorem of concave programming: its proof uses the Fritz John and Kuhn–Tucker Theorems and Theorem 1.

THEOREM 2 Any Kuhn–Tucker point for (P) is a solution vector.

PROOF Let the member $\hat{\mathbf{x}}$ of X and the multipliers $\lambda_1, \ldots, \lambda_m$ satisfy (I*) and (II*). Now (II*) states that

$$\lambda_i \geq 0, \quad f_i(\hat{\mathbf{x}}) \geq 0 \quad \text{and} \quad \lambda_i f_i(\hat{\mathbf{x}}) = 0 \text{ for } i = 1, \ldots, m.$$

Thus $\hat{\mathbf{x}}$ is feasible for (P) and

$$\sum_{i=1}^{m} \lambda_i f_i(\hat{\mathbf{x}}) = 0. \tag{1}$$

Given any $\mathbf{x}$ which is feasible for (P) we have

$$\sum_{i=1}^{m} \lambda_i f_i(\mathbf{x}) \geq 0 \tag{2}$$

by the non-negativity of the multipliers. By (I*), (1) and (2), $f_0(\hat{\mathbf{x}}) \geq f_0(\mathbf{x})$.

QED

THEOREM 3 Let the convex set X be open and let the concave functions $f_0, f_1, \ldots, f_m$ be C^1. Let $\hat{\mathbf{x}}$ be a solution vector for (P). Then $\hat{\mathbf{x}}$ is a Fritz John point. If, in addition, *any one* of the following three constraint qualifications (CQ1)–(CQ3) is satisfied, then $\hat{\mathbf{x}}$ is a Kuhn–Tucker point.

(CQ1) The gradients $\nabla f_1(\hat{\mathbf{x}}), \ldots, \nabla f_m(\hat{\mathbf{x}})$ are linearly independent.
(CQ2) The functions $f_1, \ldots, f_m$ are all affine.
(CQ3) There exists $\mathbf{x}^*$ in X such that $f_i(\mathbf{x}^*) > 0$ for $i = 1, \ldots, m$.

PROOF By the Fritz John Theorem (9.2.4), there exist multipliers ϕ_0, $\phi_1, \ldots, \phi_m$ satisfying (II), (III) and $\nabla g(\hat{\mathbf{x}}) = \mathbf{0}$, where

$$g(\mathbf{x}) = \sum_{i=0}^{m} \phi_i f_i(\mathbf{x}).$$

For each i, $\phi_i \geq 0$ and f_i is concave. Hence g is concave, and (I) follows from Theorem 1.

It remains to show that each of (CQ1), (CQ2) and (CQ3) is a constraint qualification. (CQ1) is essentially the constraint qualification of Theorem 9.3.1: see Exercise 2. (CQ2) states that all constraints are linear and is therefore the constraint qualification of Theorem 9.3.2. To prove that (CQ3) does the trick, let $\mathbf{x}^*$ be as given. If $\phi_i = 0$ for all $i \geq 1$ then $\phi_0 > 0$ by (III). Now suppose that $\phi_i > 0$ for some $i \geq 1$. Then by (II) and (CQ3),

$$\sum_{i=1}^{m} \phi_i f_i(\mathbf{x}^*) > 0 = \sum_{i=1}^{m} \phi_i f_i(\hat{\mathbf{x}}).$$

Applying (I) with $\mathbf{x} = \mathbf{x}^*$, we have $\phi_0 f_0(\hat{\mathbf{x}}) > \phi_0 f_0(\mathbf{x}^*)$. Thus $\phi_0 > 0$. QED

(CQ3) is known as the *Slater condition* of concave programming. Like (CQ2) and unlike (CQ1) it is a 'global' constraint qualification, in the sense that it does not explicitly mention $\hat{\mathbf{x}}$. Thus Theorems 2 and 3 tell us that if either (CQ2) or (CQ3) is satisfied, the set of Kuhn–Tucker points for (P) is exactly the set of solution vectors.

So far in this section we have considered only programmes where all explicit constraints are weak inequalities. The case of equation constraints requires some care. Of course, any constraint $f_i(\mathbf{x}) = 0$ can be written as two constraints $f_i(\mathbf{x}) \geq 0$ and $-f_i(\mathbf{x}) \geq 0$. But if both these are to fit into the framework of concave programming then f_i must be both concave and convex, and therefore affine by Theorem 10.2.10. Thus *linear* equation constraints may be brought into concave programming without much difficulty, while any other form of equation constraint takes us outside the realm of concave programming.

'Without much difficulty', but not without any! The problem is that when we write a linear equation constraint as two weak inequalities we automatically violate the Slater condition. It is therefore useful to have a fourth constraint qualification which generalises both (CQ2) and (CQ3). This is the *Slater-affine condition*:

(CQ4) There exists $\mathbf{x}^*$ in X such that $f_i(\mathbf{x}^*) \geq 0$ for $i = 1, \ldots, m$, with equality only if f_i is affine.

It may be shown that Theorem 3 remains true when (CQ2) and (CQ3) are replaced by (CQ4). This is not proved here.

Exercises

1. A firm can produce output y using positive quantities x_1, x_2, x_3 of three inputs. The production function is

$$y = (3/x_1 + 4/x_2 + 1/x_3)^{-1/3}.$$

Notice that this is a CES production function with elasticity of substitution $\frac{1}{2}$ and diminishing returns to scale.

Suppose that the inputs can be obtained at prices

$$p_1 = 12, \quad p_2 = 9, \quad p_3 = 16,$$

and that the firm wishes to minimise total cost subject to the constraint $y \geq q$ where q is a given positive constant.

Find the optimal values of x_1, x_2, x_3 and total and marginal cost.

2. Let X, $\hat{\mathbf{x}}$ and $f_0, f_1, \ldots, f_m$ be as in Theorem 3. Let $\hat{M}$ be the set of those $i = 1, \ldots, m$ for which $f_i(\hat{\mathbf{x}}) = 0$. Show that the conclusions of the theorem remain true when (CQ1)–(CQ3) are replaced by the following.

(CQ1*) The gradients $\{\nabla f_i(\hat{\mathbf{x}}): i \in \hat{M}\}$ are linearly independent.

(CQ2*) f_i is affine for all i in $\hat{M}$.

(CQ3*) There exists $\mathbf{x}^0$ in X such that $f_i(\mathbf{x}^0) > 0$ for all i in $\hat{M}$.

Clearly (CQ1*) is somewhat weaker (more general) than (CQ1), and (CQ2*) is weaker than (CQ2). On the other hand, (CQ3*) is *equivalent* to the Slater condition (CQ3): why?

11.2 Relaxing the assumptions

The assumptions we want to relax are those in Theorem 3 of the last section, which say that X is open and $f_0, f_1, \ldots, f_m$ are C^1 functions.

Let $f_0, f_1, \ldots, f_m$ be concave functions on a convex set X in R^n and let the concave programme (P) be as in the last section. Notice that the formulation of (P) makes no assumptions about X being open or about the functions being differentiable, and nor do the definitions of Fritz John points and Kuhn–Tucker points. *Our motivation* for such definitions was in terms of gradients being zero, and our proofs of Theorem 11.1.3 and Theorem 11.1.1 used the results of Chapter 9. But this turns out to be unnecessary. The main theorem of concave programming, without differentiability assumptions, runs as follows.

THEOREM 1 Let $f_0, f_1, \ldots, f_m$ be concave functions on a convex set X in R^n, and let the problem (P) be as in Section 11.1. Then any Kuhn–Tucker point for (P) is a solution vector, and any solution vector is a Fritz John point. If there exists $\mathbf{x}^*$ in X such that

$$f_i(\mathbf{x}^*) > 0 \quad \text{for } i = 1, \ldots, m$$

then any solution vector is a Kuhn–Tucker point.

Notice that the constraint qualification of Theorem 1 is the Slater condition (CQ3), which rules out linear equation constraints. The reason for using this rather than the Slater-affine condition (CQ4) is that the latter is not a constraint qualification under the conditions of Theorem 1.

To illustrate this mildly distressing fact, we use again the example of Section 9.2, with $f_0(\mathbf{x}) = x_2$. The problem is to

$$\text{maximise } x_2 \quad \text{subject to } x_1 \geq 0 \text{ and } x_1 + x_2^2 \leq 0.$$

The only feasible point is the origin, so this is the solution vector.

This problem can be given a concave programming formulation (P_1) by setting

$$m = 2, \quad f_0(\mathbf{x}) = x_2, \quad f_1(\mathbf{x}) = x_1, \quad f_2(\mathbf{x}) = -x_1 - x_2^2$$

and $X = R^2$. By Theorem 1, $\mathbf{0}$ is a Fritz John point for (P_1). The usual argument shows that $\phi_0 = 0$. Thus the solution vector $\mathbf{0}$ is not a Kuhn–Tucker point. It is easily seen that all constraint qualifications (CQ1)–(CQ4) of the last section are violated.

Now we come to the main point. Notice that we can also formulate the given problem as a concave programme by treating $x_1 + x_2^2 \leq 0$ as an *implicit* constraint. In this formulation, which we call (P_2),

$$m = 1, \quad f_0(\mathbf{x}) = x_2, \quad f_1(\mathbf{x}) = x_1$$

and X is the set $\{\mathbf{x} \in R^2\colon x_1 + x_2^2 \leq 0\}$. A diagram easily demonstrates that the problem

$$\text{maximise } x_2 + \lambda x_1 \quad \text{subject to } \mathbf{x} \in X$$

does not have a solution at the origin for any non-negative λ. Thus the solution vector $\mathbf{0}$ to (P_2) is not a Kuhn–Tucker point. But the only explicit constraint in (P_2) is linear, so (CQ4) is satisfied.

What has gone wrong in this example is that the only feasible vector for (P_2) lies on the boundary of X. This suggests that one way to bring back the Slater-affine constraint qualification is to insist that X be open. But this is actually unnecessary: to state what is required, we introduce the notion of an internal point of a convex set.

Let X be convex and let $\mathbf{x} \in X$. We say that $\mathbf{x}$ is an *internal point* of X if, for any $\mathbf{y}$ in X, the line segment from $\mathbf{y}$ to $\mathbf{x}$ can be extended beyond $\mathbf{x}$ without passing outside X. Thus $\mathbf{x}$ is an internal point of X if, for any $\mathbf{y}$ in X, there exists a positive number δ depending on $\mathbf{y}$ such that

$$(1 + \delta)\mathbf{x} - \delta\mathbf{y} \in X.$$

The Slater-affine condition for general concave programming is as follows.

(CQ5) There exists an internal point $\mathbf{x}^*$ of X such that $f_i(\mathbf{x}^*) \geq 0$ for $i = 1, \ldots, m$, with equality only if f_i is affine

The following theorem states that this is a constraint qualification for (P).

THEOREM 2 Let $f_0, f_1, \ldots, f_m$ be concave functions on a convex set X in R^n, and let (P) be as in Section 11.1. If (CQ5) is satisfied, any solution vector for (P) is a Kuhn–Tucker point.

The importance of Theorem 2 is that it allows linear equation constraints. In fact, the Slater condition (CQ3) is a special case of (CQ5): for if there exists $\mathbf{x}^*$ in X such that $f_i(\mathbf{x}^*) > 0$ for $i = 1, \ldots, m$, there exists an internal point $\mathbf{x}^{**}$ with the same property. The reader is asked to prove this in Exercises 2–5.

Exercises

1. A consumer consumes non-negative quantities of three goods x_1, x_2, x_3 available at prices 3, 2, 6, respectively. His income is 12 and his utility function is

$$2x_1 + 5\sqrt{x_2 x_3}.$$

What quantities x_1, x_2, x_3 maximise utility subject to the budget constraint?

2. Let X be a non-empty convex set in R^n, and let L be the set of all points of the form $\lambda(\mathbf{x} - \mathbf{y})$ where $\mathbf{x}$ and $\mathbf{y}$ are any members of X and λ is any scalar. Show that L is a linear subspace of R^n.

The set L is the linear subspace *parallel* to X: illustrate this with a diagram in R^2. The dimension of L is called the dimension of X and denoted by dim X; this definition of the dimension of a convex set does not lead to ambiguities when X is a linear subspace, for then $L = X$. Notice that dim $X = 0$ if and only if X is a singleton.

3. Points $\mathbf{x}_0, \mathbf{x}_1, \ldots, \mathbf{x}_k$ of a convex set X in R^n are said to be *in general position* if the $(n+1)$-vectors

$$(\mathbf{x}_0\colon 1), \quad (\mathbf{x}_1\colon 1), \ldots, \quad (\mathbf{x}_k\colon 1)$$

are linearly independent.

Let X be a non-empty convex set of dimension d. Prove:

(i) for any $\mathbf{x}$ in X one may choose $1 + d$ points in general position in X, one of which is $\mathbf{x}$;

(ii) if $\mathbf{x}_0, \mathbf{x}_1, \ldots, \mathbf{x}_d$ are $1 + d$ points in general position in X, then the point

$$(1+d)^{-1}(\mathbf{x}_0 + \mathbf{x}_1 + \cdots + \mathbf{x}_d)$$

is an internal point of X.

The set of all internal points of a convex set X is called the *relative interior* of X. It follows from (i) and (ii) that any non-empty convex set has a non-empty relative interior.

4. Let $\mathbf{x}$ and $\mathbf{y}$ be members of a convex set X in R^n, with $\mathbf{x}$ internal. Show that if $0 < \alpha < 1$ then $\alpha\mathbf{x} + (1 - \alpha)\mathbf{y}$ is an internal point of X.

5. Use the results of Exercises 3 and 4 to prove the last assertion of this section.

11.3 Duality

As in the last two sections, let $f_0, f_1, \ldots, f_m$ be concave functions on a convex set X in R^n. We are interested in the problem

(P) maximise $f_0(\mathbf{x})$ subject to $f_i(\mathbf{x}) \geq 0$ for $i = 1, \ldots, m$.

We begin by rephrasing the definition of a Kuhn–Tucker point. The point $\hat{\mathbf{x}}$ of X is a Kuhn–Tucker point for (P) if and only if there exist multipliers $\lambda_1, \ldots, \lambda_m$ such that

(I^0) $f_0(\hat{\mathbf{x}}) \geq f_0(\mathbf{x}) + \sum_{i=1}^{m} \lambda_i f_i(\mathbf{x})$ for all $\mathbf{x}$ in X,

(II^0) λ_i and $f_i(\hat{\mathbf{x}})$ are non-negative for $i = 1, \ldots, m$.

It is left to the reader to demonstrate that these conditions, taken together, are equivalent to (I*) and (II*) of Section 11.1. This result implies the following proposition, which is easy to prove: if there exists a Kuhn–Tucker point for (P) with multipliers $\lambda_1, \ldots, \lambda_m$, then any solution vector is a Kuhn–Tucker point with the same multipliers. This means that we can speak of multipliers being associated with a *programme* (P) rather than a particular solution vector.

A soluble programme (P) is said to be *stable* if one of its solution vectors is a Kuhn–Tucker point, in which case all are. Theorem 11.2.2 states that any soluble programme which satisfies the constraint qualifications (CQ5) is stable. The motivation for the term 'stable' comes from sensitivity analysis: it may be shown that any concave programme which is *not* stable may be rendered infeasible by an artibrarily small perturbation of the constraints. (We should, however, remark that some stable programmes also have this property, so the term is slightly misleading.)

Suppose that (P) is soluble, let its solution value be v and let $\hat{\mathbf{x}}$ be a solution vector. Suppose also that (P) is stable and let $\hat{\mathbf{p}}$ be an m-vector of Kuhn–Tucker multipliers. Then (I^0) and (II^0) hold with $\lambda_i = \hat{p}_i$ for $i = 1, \ldots, m$.

For any $\mathbf{x}$ in X, and *any* non-negative m-vector $\mathbf{p}$, we consider the *Lagrangian*

$$L(\mathbf{x}, \mathbf{p}) = f_0(\mathbf{x}) + \sum_{i=1}^{m} p_i f_i(\mathbf{x})$$

as a function of $\mathbf{x}$ and $\mathbf{p}$. By our choice of $\hat{\mathbf{x}}$ and $\hat{\mathbf{p}}$,

$$v = f_0(\hat{\mathbf{x}}) = L(\hat{\mathbf{x}}, \hat{\mathbf{p}}) \tag{1}$$

and $v \geq L(\mathbf{x}, \hat{\mathbf{p}})$ for any $\mathbf{x}$ in X. Also $\hat{\mathbf{x}}$ is feasible for (P) so $L(\hat{\mathbf{x}}, \mathbf{p}) \geq$

$f_0(\hat{\mathbf{x}})$ for any non-negative $\mathbf{p}$. Thus

$$L(\mathbf{x}, \hat{\mathbf{p}}) \leq L(\hat{\mathbf{x}}, \hat{\mathbf{p}}) \leq L(\hat{\mathbf{x}}, \mathbf{p}) \tag{2}$$

for all $\mathbf{x}$ in X and all $\mathbf{p} \geq \mathbf{0}$. This is called the *saddle point property* of concave programming and has applications to the theory of economic planning (see Heal, 1973, Chapter 4).

To get from the saddle point property to duality we fix an arbitrary $\mathbf{p}$ in R^m_+ and consider the problem

$$\mathrm{F}(\mathbf{p}): \quad \text{maximise } L(\mathbf{x}, \mathbf{p}) \text{ subject to } \mathbf{x} \in X.$$

If this problem has a solution we denote the solution value by $\phi(\mathbf{p})$; if it is unbounded let $\phi(\mathbf{p}) = \infty$. (We know from the remarks at the end of Section 5.2 that a nonlinear maximisation problem may be feasible and bounded without being soluble. If $\mathrm{F}(\mathbf{p})$ has this property, $\phi(\mathbf{p})$ is taken as 'least upper bound'.)

By (1) and the right-hand inequality in (2), $\phi(\mathbf{p}) \geq v$ for all $\mathbf{p} \geq \mathbf{0}$, with equality if $\mathbf{p} = \hat{\mathbf{p}}$. Thus the problem

$$(\mathrm{P}^*) \quad \text{minimise } \phi(\mathbf{p}) \quad \text{subject to } \mathbf{p} \geq \mathbf{0}$$

has solution vector $\hat{\mathbf{p}}$ and solution value v.

Now (P*) is the *dual* problem to the concave programme (P): it is a minimisation problem in m-space with the same solution value as (P), and any vector of Kuhn–Tucker multipliers for (P) is a solution to it. In other words, (P*) has the properties we have come to associate with a dual problem.

It also reduces to the usual notion of dual in the case of linear programming. To see this, put the standard-form linear programme (S) into the form (P) by taking $\mathbf{x} \geq \mathbf{0}$ as the implicit constraint and $\mathbf{b} - \mathbf{A}\mathbf{x} \geq \mathbf{0}$ as the explicit one. Then

$$L(\mathbf{x}, \mathbf{p}) = \mathbf{c}^{\mathrm{T}}\mathbf{x} + \mathbf{p}^{\mathrm{T}}(\mathbf{b} - \mathbf{A}\mathbf{x}) = \mathbf{b}^{\mathrm{T}}\mathbf{p} + \mathbf{x}^{\mathrm{T}}(\mathbf{c} - \mathbf{A}^{\mathrm{T}}\mathbf{p}).$$

Fix $\mathbf{p} \geq \mathbf{0}$ and consider $\mathrm{F}(\mathbf{p})$. If $\mathbf{c} \leq \mathbf{A}^{\mathrm{T}}\mathbf{p}$ then $L(\mathbf{x}, \mathbf{p}) \leq \mathbf{b}^{\mathrm{T}}\mathbf{p}$ with equality if $\mathbf{x} = \mathbf{0}$. If $\mathbf{c} - \mathbf{A}^{\mathrm{T}}\mathbf{p}$ has positive component k, we can make $L(\mathbf{x}, \mathbf{p})$ as large as we like by making x_k large and positive. Thus $\phi(\mathbf{p})$ is $\mathbf{b}^{\mathrm{T}}\mathbf{p}$ if $\mathbf{A}^{\mathrm{T}}\mathbf{p} \geq \mathbf{c}$ and ∞ otherwise. Thus in this case the rather unconstrained looking problem (P*) does indeed reduce to

$$(\mathrm{D}) \quad \text{minimise } \mathbf{b}^{\mathrm{T}}\mathbf{p} \quad \text{subject to } \mathbf{A}^{\mathrm{T}}\mathbf{p} > \mathbf{c} \text{ and } \mathbf{p} \geq \mathbf{0}.$$

Having accepted all this, the reader may still find the dual construction (P*) rather artificial. It certainly doesn't look as if it has the neat two-way relationship with (P) that (D) has with (S), and we have not stated any theorems to that effect.

This suggests that the (P) formulation of concave programmes is not the ideal one for making the relation between a programme and its dual really transparent. Other points of departure have been suggested, one nice one being described in Cass (1974). But a full discussion of these issues requires another book.

Appendix A: Bland's Refinement Avoids Cycling

Bland's Refinement of the Simplex Method is a technique for avoiding cycling. In the notation of Section 6.2 it consists of choosing among admissible pivot columns, and among admissible pivot rows, by selecting in each case the one with the smallest label. Let us consider this as two rules, one (C) for choosing among columns and the other (R) for choosing among rows, given the choice of pivot column. This appendix gives a version of Bland's proof that the procedure eliminates the possibility of cycling.

Suppose we are attempting to solve programmes (S) and (D) by the simplex method, that Rule C has been followed at each step, and that cycling occurs. We must show that Rule R has been violated at some stage.

Let T be the set of integers which appear both as row labels and as column labels in the course of the cycle. The hypothesis of cycling implies that the left-border entry of the pivot row is zero for each scheme of the cycle. Thus the solution to $\mathbf{Ax} + \mathbf{y} = \mathbf{b}$ read off in the usual manner from each scheme of the cycle remains unchanged from scheme to scheme. Hence:

(*) for any scheme of the cycle, each row of that scheme whose label is in T has left-border entry zero.

Let τ be the *largest* member of T. Since τ is in T, we may choose schemes Σ^α, Σ^β in the cycle such that τ labels the pivot *column* of Σ^α and the pivot *row* of Σ^β. In Σ^α denote entries by α's, labels by π's and the pivot entry by α_{rs}. In Σ^β denote entries by β's, labels by ψ's and the pivot entry by β_{tu}. Thus $\pi_s = \psi_{n+t} = \tau$.

Let the $(n+m)$-vector $(\mathbf{q}: \mathbf{p}) = \mathbf{w}$ be the solution to $\mathbf{A}^{\mathrm{T}}\mathbf{p} - \mathbf{q} = \mathbf{c}$ read off in the usual manner from the top row of Σ^{α}. Also, let the $(n+m)$-vector $(\mathbf{x}: \mathbf{y}) = \mathbf{z}$ be the solution to $\mathbf{Ax} + \mathbf{y} = \mathbf{b}$ read off from the row system of Σ^{β} in the following (unusual) manner:

$$\begin{aligned} &z_k = -1 \quad \text{if } k = \psi_u, \\ &z_k = \beta_{i0} + \beta_{iu} \quad \text{if } k = \psi_{n+i} \quad (i = 1, \ldots, m), \\ &z_k = 0 \text{ otherwise.} \end{aligned}$$

Then $\mathbf{b}^{\mathrm{T}}\mathbf{p} = \alpha_{00}$ and $\mathbf{c}^{\mathrm{T}}\mathbf{x} = \beta_{00} + \beta_{0u}$, so

$$\begin{aligned} \mathbf{w}^{\mathrm{T}}\mathbf{z} &= (\mathbf{A}^{\mathrm{T}}\mathbf{p} - \mathbf{c})^{\mathrm{T}}\mathbf{x} + \mathbf{p}^{\mathrm{T}}(\mathbf{b} - \mathbf{Ax}) = \mathbf{b}^{\mathrm{T}}\mathbf{p} - \mathbf{c}^{\mathrm{T}}\mathbf{x} \\ &= \alpha_{00} - \beta_{00} - \beta_{0u}. \end{aligned}$$

Since Σ^{α} and Σ^{β} are in the cycle, $\alpha_{00} = \beta_{00}$. Since β_{tu} is the pivot entry in Σ^{β}, $\beta_{0u} < 0$. Thus $\mathbf{w}^{\mathrm{T}}\mathbf{z} > 0$.

Let ξ denote the sum of the terms $\alpha_{ij}(\beta_{i0} + \beta_{iu})$ over all pairs (i, j) such that column j of Σ^{α} has the same label as row i of Σ^{β}. Also let

$$\eta = \begin{cases} 0 & \text{if } \psi_u \text{ is a row label in } \Sigma^{\alpha}, \\ \alpha_{0k} & \text{if } \psi_u \text{ labels column } k \text{ of } \Sigma^{\alpha}. \end{cases}$$

By construction of $\mathbf{w}$ and $\mathbf{z}$, $\mathbf{w}^{\mathrm{T}}\mathbf{z} = \xi - \eta$. Hence $\xi > \eta$.

We now show that $\eta \geq 0$. It suffices to consider the case where $\eta = \alpha_{0k}$ for some k. Then $\pi_k = \psi_u$, so τ and π_k label respectively the pivot row and pivot column of Σ^{β}. Thus π_k is a member of T other than, *and therefore less than*, τ. But the pivot column of Σ^{α} is s (with label τ) rather than k (with label $\pi_k < \tau$) and Rule C is observed. Thus column k of Σ^{α} is not an admissible pivot column, so $\eta = \alpha_{0k} \geq 0$.

Since $\xi > \eta \geq 0$ we can choose a member σ of T which labels a column (say j) of Σ^{α} and a row (say i) of Σ^{β}, such that $\alpha_{0j}(\beta_{i0} + \beta_{iu}) > 0$. By (*), $\beta_{i0} = 0$: thus α_{0j} and β_{iu} are either both positive or both negative. Since α_{is} and β_{tu} are the pivot entries in Σ^{α} and Σ^{β} respectively, $\alpha_{0s} < 0 < \beta_{tu}$. Thus the pair (i, j) is not the same as the pair (t, s): this means that σ is a member of T other than, and therefore less than, τ. Now the pivot column of Σ^{α} is s (with label τ) rather than j (with label $\sigma < \tau$) and Rule C is observed: hence $\alpha_{0j} \geq 0$. But we have already shown that α_{0j} and β_{iu} are non-zero numbers of the same sign and that $\beta_{i0} = 0$. Hence $\beta_{i0} = 0 < \beta_{iu}$, which means that row i (with label $\sigma < \tau$) is an admissible pivot row for Σ^{β}. Thus the choice of β_{tu} rather than β_{iu} as pivot entry in Σ^{β} violates Rule R.

Appendix B: The Marginal Value Theorem

The object of this appendix is to prove the Theorem of Section 7.3.

$\mathbf{A}$ is an $m \times n$ matrix, $\mathbf{b}$ and $\mathbf{d}$ are m-vectors and $\mathbf{c}$ is an n-vector. The linear programmes (S) and (D) have their usual meanings and we assume that they are soluble. For each positive ε we are interested in the linear programme

$$(\text{S}_\varepsilon) \quad \text{maximise } \mathbf{c}^\text{T}\mathbf{x} \quad \text{subject to } \mathbf{A}\mathbf{x} \leq \mathbf{b} + \varepsilon\mathbf{d} \text{ and } \mathbf{x} \geq \mathbf{0}$$

and its dual

$$(\text{D}_\varepsilon) \quad \text{minimise } (\mathbf{b} + \varepsilon\mathbf{d})^\text{T}\mathbf{p} \quad \text{subject to } \mathbf{A}^\text{T}\mathbf{p} \geq \mathbf{c} \text{ and } \mathbf{p} \geq \mathbf{0}.$$

We are also interested in the linear programme

$$\begin{aligned}(\text{M}) \quad & \text{minimise } \mathbf{d}^\text{T}\mathbf{p} \quad \text{subject to} \\ & \mathbf{p} \text{ being a solution vector for (D).}\end{aligned}$$

This may be written

$$\begin{aligned}(\text{M}) \quad & \text{minimise } \mathbf{d}^\text{T}\mathbf{p} \text{ subject to} \\ & \mathbf{A}^\text{T}\mathbf{p} \geq \mathbf{c},\ -\mathbf{b}^\text{T}\mathbf{p} \geq -f(0) \text{ and } \mathbf{p} \geq \mathbf{0}\end{aligned}$$

where $f(0)$ denotes the common solution value of (S) and (D).

Notice that (M) is the dual of the following standard-form linear programme in $n+1$ variables $x_1, \ldots, x_n, x_*$:

$$\begin{aligned}(\text{M}^*) \quad & \text{maximise } \mathbf{c}^\text{T}\mathbf{x} - f(0)x_* \quad \text{subject to} \\ & \mathbf{A}\mathbf{x} - x_*\mathbf{b} \leq \mathbf{d},\ \mathbf{x} \geq \mathbf{0} \text{ and } x_* \geq 0.\end{aligned}$$

The statement of the Marginal Value Theorem in the text is periphrastic, and we can straighten it out by using the extended real number system. For each positive ε we define $f(\varepsilon)$ to be the solution value of (S_ε) if that programme is soluble, and $-\infty$ if (S_ε) is infeasible. We define the extended real number λ to be the solution value of (M) if (M) is soluble, $-\infty$ if (M) is feasible and unbounded. Recall from the text that these are the only possible cases. The theorem can then be given the following simple form.

THEOREM $f(\varepsilon) \leq f(0) + \varepsilon\lambda$ for all positive ε, with equality for all sufficiently small positive ε.

PROOF We first prove the weak-inequality part. Choose a positive ε. If $f(\varepsilon) = -\infty$ there is nothing to prove, so suppose that (S_ε) and (D_ε) are soluble with common solution value $f(\varepsilon)$. Let $\mathbf{p}$ be a solution vector for (D). Then $\mathbf{b}^{\mathrm{T}}\mathbf{p} = f(0)$: also $\mathbf{p}$ is feasible for (D_ε) so $(\mathbf{b} + \varepsilon\mathbf{d})^{\mathrm{T}}\mathbf{p} \geq f(\varepsilon)$. But then

$$\varepsilon\mathbf{d}^{\mathrm{T}}\mathbf{p} \geq f(\varepsilon) - f(0).$$

Since this is so for every $\mathbf{p}$ which is feasible for (M), the feasible programme (M) is bounded and $\varepsilon\lambda \geq f(\varepsilon) - f(0)$.

We have now shown that $f(\varepsilon) \leq f(0) + \varepsilon\lambda$ for all positive ε. In particular, if $\lambda = -\infty$ then

$$f(\varepsilon) = f(0) + \varepsilon\lambda = -\infty \quad \text{for all positive } \varepsilon.$$

So from now on assume that λ is finite: we want to show that for small ε, $f(\varepsilon) \geq f(0) + \varepsilon\lambda$. Thus it suffices to prove that for all sufficiently small positive ε there exists an n-vector $\mathbf{x}(\varepsilon)$, feasible for (S_ε), such that

$$\mathbf{c}^{\mathrm{T}}\mathbf{x}(\varepsilon) = f(0) + \varepsilon\lambda.$$

Since (S) is assumed to be soluble we may choose a solution vector $\mathbf{x}^0$ for (S). Then

$$\mathbf{A}\mathbf{x}^0 \leq \mathbf{b}, \quad \mathbf{x}^0 \geq 0 \quad \text{and} \quad \mathbf{c}^{\mathrm{T}}\mathbf{x}^0 = f(0). \tag{1}$$

Since λ is assumed to be finite we may apply the Duality Theorem to (M) and (M*), inferring that λ is the solution value for (M*). Let $\mathbf{x} = \mathbf{x}^1$, $x_* = \theta$ be a solution to (M*). Then

$$\mathbf{A}\mathbf{x}^1 \leq \mathbf{d} + \theta\mathbf{b}, \quad \mathbf{x}^1 \geq \mathbf{0}, \quad \theta \geq 0 \quad \text{and} \quad \mathbf{c}^{\mathrm{T}}\mathbf{x}^1 - f(0)\theta = \lambda. \tag{2}$$

By 'sufficiently small positive ε' we mean a number ε such that

$$\varepsilon > 0 \quad \text{and} \quad \varepsilon\theta \leq 1. \tag{3}$$

So if $\theta = 0$ any positive ε will do; if $\theta > 0$ we require that $\varepsilon \leq 1/\theta$. For such ε let

$$\mathbf{x}(\varepsilon) = (1 - \varepsilon\theta)\mathbf{x}^0 + \varepsilon\mathbf{x}^1.$$

Then $\mathbf{x}(\varepsilon) \geq \mathbf{0}$. By (1), (2) and (3)

$$\begin{aligned}\mathbf{Ax}(\varepsilon) &= (1 - \varepsilon\theta)\mathbf{Ax}^0 + \varepsilon\mathbf{Ax}^1 \\ &\leq (1 - \varepsilon\theta)\mathbf{b} + \varepsilon(\mathbf{d} + \theta\mathbf{b}) \\ &= \mathbf{b} + \varepsilon\mathbf{d}.\end{aligned}$$

Thus $\mathbf{x}(\varepsilon)$ is feasible for (S_ε). Also by (1) and (2),

$$\begin{aligned}\mathbf{c}^{\mathrm{T}}\mathbf{x}(\varepsilon) &= (1 - \varepsilon\theta)\mathbf{c}^{\mathrm{T}}\mathbf{x}^0 + \varepsilon\mathbf{c}^{\mathrm{T}}\mathbf{x}^1 \\ &= (1 - \varepsilon\theta)f(0) + \varepsilon(\lambda + \theta f(0)) \\ &= f(0) + \varepsilon\lambda.\end{aligned}$$

QED

Appendix C: Determinants

The only reference we have made to determinants in this book was in Section 10.2, in connection with testing functions for concavity. We did not mention them in Chapter 2 because we were too busy proving theorems and doing sums; and determinants are quite unnecessary for proving the main results about systems of linear equations and do not provide a helpful method for the numerical solution of such systems.

In economic theory, one often wants to solve a system of linear equations whose coefficients are symbols rather than numbers. For this purpose, one *sometimes* needs explicit formulae for matrix inversion and for the solution to a nonsingular system. The need for this arises far less often than many economists think. But the literature exists, and such formulae involve determinants. Hence this appendix.

Another reason for knowing about determinants is that in more advanced mathematics (multiple integration, eigenvalue theory), determinants really are necessary for proving theorems and doing sums. The concavity test is related to eigenvalue theory.

In this appendix we outline the theory of determinants, omitting most of the proofs. For a much more detailed treatment see Chapter 4 of Strang (1976).

We begin with some definitions concerning permutations. A *permutation* of length n is a row vector consisting of the integers $1, \ldots, n$ either in their natural order or in some other order (recall that the labelling procedure in pivot-and-switch uses permutations of length $n + m$). Let $\pi = (\pi_1, \ldots, \pi_n)$ be a permutation of length n and let i and

j be integers such that $1 \le i < j \le n$: we say that the pair (i, j) is an *inversion* of π if $\pi_i > \pi_j$. A permutation is said to be *even* if it has an even number of inversions, *odd* if it has an odd number. The *sign* $s(\pi)$ of a permutation π is defined to be 1 if π is even, -1 if π is odd.

To illustrate this, let $n = 4$ and consider the permutations

$$\pi^1 = (1, 2, 3, 4), \quad \pi^2 = (2, 1, 4, 3), \quad \pi^3 = (2, 4, 1, 3).$$

π^1 has no inversions and is therefore even (zero is an even number). The inversions of π^2 are (1, 2) and (3, 4) so $s(\pi^2) = 1$. The inversions of π^3 are (1, 3), (2, 3) and (2, 4) so $s(\pi^3) = -1$.

The *determinant* of the square matrix $\mathbf{A}$ of order n is the number

$$\sum_{\pi} s(\pi) a_{1\pi_1} a_{2\pi_2} \cdots a_{n\pi_n},$$

where the sum is taken over all permutations of length n. Notice that only square matrices have determinants; most of this appendix will be concerned with square matrices (the exception is the discussion of rank in and around Theorem 8). The determinant of $\mathbf{A}$ is written det $\mathbf{A}$ or

$$\begin{vmatrix} a_{11} & \cdots & a_{1n} \\ & \vdots & \\ a_{n1} & \cdots & a_{nn} \end{vmatrix}.$$

The expression $|\mathbf{A}|$ for det $\mathbf{A}$ is used by some authors, but this tends to lead to confusion with absolute values and norms.

With some effort, it is possible to derive the following two important properties of determinants directly from the definition.

THEOREM 1 $\det \mathbf{A}^{\mathrm{T}} = \det \mathbf{A}$.

THEOREM 2 $\det (\mathbf{AB}) = (\det \mathbf{A})(\det \mathbf{B})$.

We turn now to the calculation of determinants. There are three cases where it is easy to calculate det $\mathbf{A}$ directly from the definition. Case 1 is where $n = 1$: $\det(\alpha) = \alpha$. Case 2 is where $n = 2$: here we have two permutations, the even permutation (1, 2) and the odd permutation (2, 1). Thus in this case

$$\det \mathbf{A} = \alpha_{11} a_{22} - a_{12} a_{21}.$$

Case 3 is where $\mathbf{A}$ is triangular. Let π be any permutation of length n other than the natural order. If $\mathbf{A}$ is either upper or lower triangular, at least one of the terms

$$a_{1\pi_1}, a_{2\pi_2}, \ldots, a_{n\pi_n}$$

must vanish. Thus if **A** is triangular,

$$\det \mathbf{A} = a_{11}a_{22} \cdots a_{nn}.$$

In more general cases there are two standard ways of calculating determinants. The first uses Case 3 and Gaussian elimination.

THEOREM 3 Let the square matrix **A** be reduced to the upper triangular matrix **U** by Gaussian elimination. Suppose that this involves k row interchanges. Then

$$\det \mathbf{A} = (-1)^k \det \mathbf{U}.$$

Example 1 Find det **A** where

$$\mathbf{A} = \begin{pmatrix} 1 & 3 & 2 & 5 \\ 0 & 0 & 3 & 2 \\ 1 & 5 & 4 & 0 \\ 1 & 2 & 1 & 1 \end{pmatrix}.$$

Gaussian elimination, with *one* row interchange, yields the upper triangular matrix

$$\mathbf{U} = \begin{pmatrix} 1 & 3 & 2 & 5 \\ 0 & 2 & 2 & -5 \\ 0 & 0 & 3 & 2 \\ 0 & 0 & 0 & -\frac{13}{2} \end{pmatrix}.$$

Since **U** is triangular, its determinant is the product of its diagonal entries: det **U** $= -39$. Since $k = 1$, det **A** $= +39$.

The following theorem follows immediately from Theorem 3 and the theory of Chapter 2.

THEOREM 4 The square matrix **A** is singular if and only if det **A** $= 0$.

The other way of calculating determinants is the *cofactor method*. This expresses the determinant of a square matrix of order n in terms of determinants of square matrices of order $n-1$. Iterating, we obtain an expression for an arbitrary determinant in terms of determinants of order 2, and apply Case 2 above.

Explanation of the cofactor method requires some more definitions and another theorem. Let $\mathbf{A}$ be a square matrix of order n. For any pair (i, j) we let $\mathbf{A}(i, j)$ denote the square matrix of order $n-1$ obtained by deleting row i and column j of $\mathbf{A}$. The (i, j) *cofactor* of $\mathbf{A}$ is the scalar

$$\tilde{a}_{ij} = (-1)^{i+j} \det \mathbf{A}(i, j).$$

The $n \times n$ matrix $(\tilde{a}_{ij})$ is called the *cofactor matrix* of $\mathbf{A}$ and denoted by $\tilde{\mathbf{A}}$. Its transpose $\tilde{\mathbf{A}}^{\mathrm{T}}$ is called the *adjugate* (or *adjoint*) of $\mathbf{A}$ and is denoted by adj $\mathbf{A}$.

We state the next theorem without proof.

THEOREM 5 For any square matrix $\mathbf{A}$,

$$\mathbf{A}(\operatorname{adj} \mathbf{A}) = (\det \mathbf{A})\mathbf{I} = (\operatorname{adj} \mathbf{A})\mathbf{A}. \tag{1}$$

Notice that the brackets in the left and right terms of (1) denote matrix multiplication, while the brackets in the middle term denote multiplication of the identity matrix $\mathbf{I}$ by the scalar det $\mathbf{A}$. Such ambiguities of notation will recur, and we shall not draw attention to them again.

To obtain formulae for calculating determinants, we unravel (1). Equating diagonal entries in the left-hand equation we see that

$$\det \mathbf{A} = \sum_{j=1}^{n} a_{ij}\tilde{a}_{ij} \tag{2}$$

for $i = 1, \ldots, n$. This gives us n different ways of expressing the determinant of order n in terms of determinants of order $n-1$. Another n ways can be obtained by equating diagonal entries in the right-hand equation of (1):

$$\det \mathbf{A} = \sum_{i=1}^{n} a_{ij}\tilde{a}_{ij} \tag{3}$$

for $j = 1, \ldots, n$.

The formula (2) is known as *expansion by row i*, while (3) is known as *expansion by column j*. The fact that we can expand either by a row or by a column should not be surprising in view of Theorem 1. Equating off-diagonal entries in (1) gives the *alien cofactors formula:*

$$\sum_{j=1}^{n} a_{rj}\tilde{a}_{sj} = 0 = \sum_{i=1}^{n} a_{is}\tilde{a}_{ir} \quad \text{when } r \neq s.$$

Example 2 Evaluate det **A** by the cofactor method, where **A** is as in Example 1.

It saves effort to expand by a row or column with several zeroes in it. Expanding by the second row we have

$$\det \mathbf{A} = 3\tilde{a}_{23} + 2\tilde{a}_{24}.$$

Since $(-1)^{2+3} = -1$ and $(-1)^{2+4} = +1$ we have

$$\det \mathbf{A} = -3\lambda + 2\mu$$

where

$$\lambda = \begin{vmatrix} 1 & 3 & 5 \\ 1 & 5 & 0 \\ 1 & 2 & 1 \end{vmatrix} \quad \text{and} \quad \mu = \begin{vmatrix} 1 & 3 & 2 \\ 1 & 5 & 4 \\ 1 & 2 & 1 \end{vmatrix}.$$

Expanding λ by the third column we have

$$\lambda = 5\cdot(-1)^{1+3}\cdot\begin{vmatrix} 1 & 5 \\ 1 & 2 \end{vmatrix} + 1\cdot(-1)^{3+3}\cdot\begin{vmatrix} 1 & 3 \\ 1 & 5 \end{vmatrix}$$

$$= 5(2-5) + (5-3) = -13.$$

Expanding μ by the first column we have

$$\mu = \begin{vmatrix} 5 & 4 \\ 2 & 1 \end{vmatrix} - \begin{vmatrix} 3 & 2 \\ 2 & 1 \end{vmatrix} + \begin{vmatrix} 3 & 2 \\ 5 & 4 \end{vmatrix} = -3 + 1 + 2 = 0.$$

Thus $\det \mathbf{A} = -3\lambda + 0 = 39$.

The inversion formula, which generalises Theorem 2.2.6, is given in the next theorem.

THEOREM 6 If **A** is nonsingular then $\mathbf{A}^{-1} = (\det \mathbf{A})^{-1} \operatorname{adj} \mathbf{A}$.
PROOF Let **A** be nonsingular and let $\delta = \det \mathbf{A}$. By Theorem 4, $\delta \neq 0$. Let $\mathbf{B} = \delta^{-1} \operatorname{adj} \mathbf{A}$. Multiplying (1) by δ^{-1} we have $\mathbf{AB} = \mathbf{BA} = \mathbf{I}$. QED

Theorem 6 gives a method for matrix inversion called *inversion by cofactors*. This is generally inefficient in numerical work when $n > 2$. An equally inefficient way of solving a system of linear equations is given by the next theorem.

THEOREM 7 (Cramer's Rule)
Let **A** be a nonsingular matrix of order n, **b** an n-vector. For each

$j = 1, \ldots, n$, let $\mathbf{B}_j$ be the square matrix whose jth column is $\mathbf{b}$ and whose other columns are as in $\mathbf{A}$.

Then the solution to the system $\mathbf{Ax} = \mathbf{b}$ is

$$x_j = \det \mathbf{B}_j / \det \mathbf{A} \quad (j = 1, \ldots, n).$$

PROOF Let $\mathbf{z}$ be the n-vector with components $\det \mathbf{B}_1, \ldots, \det \mathbf{B}_n$. We want to prove that $\mathbf{A}^{-1}\mathbf{b} = (\det \mathbf{A})^{-1}\mathbf{z}$.

Fix an integer k and expand $\det \mathbf{B}_k$ by column k. Since $\mathbf{B}_k$ differs from $\mathbf{A}$ only in column k, the two matrices have the same (i, k) cofactor for all i. Thus

$$z_k = b_1 \tilde{a}_{1k} + \cdots + b_k \tilde{a}_{nk}.$$

Since this is so for all k, $\mathbf{z}^{\mathrm{T}} = \mathbf{b}^{\mathrm{T}}\tilde{\mathbf{A}}$. Transposing, $\mathbf{z} = (\operatorname{adj} \mathbf{A})\mathbf{b}$. Multiplying by $(\det \mathbf{A})^{-1}$ and applying Theorem 6, we obtain the required result. QED

Recall from Section 2.4 that a submatrix of a matrix $\mathbf{A}$, not necessarily square, is a matrix obtained from $\mathbf{A}$ by deleting some (or none) of its columns and some (or none) of its rows. A *minor* of a matrix is the determinant of one of its square submatrices. By Theorem 4, we may rephrase Theorem 2.4.2 as follows.

THEOREM 8 The rank of a matrix is the maximal order of its non-vanishing minors.

Notice that Theorem 8 can be used as a definition of the rank of a matrix that does not invoke the idea of linear independence. Similarly, Theorem 4 may be used to define a singular matrix without mentioning linear dependence. These definitions were the natural ones from the point of view of those nineteenth-century mathematicians who were more interested in determinants for their own sake than in systems of linear equations.

Returning to square matrices we define a *principal submatrix* of an $n \times n$ matrix $\mathbf{A}$ to be a submatrix obtained from $\mathbf{A}$ by the following rule: row k is deleted if and only if column k is deleted. A *leading principal submatrix* of $\mathbf{A}$ is obtained when the deleted rows and columns are the last m rows and columns for some $m = 0, 1, \ldots, n - 1$. Thus the leading principal submatrices of the 3×3 matrix

$$\mathbf{A} = \begin{pmatrix} 2 & 1 & 5 \\ 6 & 7 & 9 \\ 3 & 4 & 8 \end{pmatrix}$$

are **A** itself, the 2×2 matrix $\begin{pmatrix} 2 & 1 \\ 6 & 7 \end{pmatrix}$ and the scalar 2.

The other principal submatrices are the other two diagonal entries and the 2×2 matrices

$$\begin{pmatrix} 2 & 5 \\ 3 & 8 \end{pmatrix} \quad \text{and} \quad \begin{pmatrix} 7 & 9 \\ 4 & 8 \end{pmatrix}.$$

A (*leading*) *principal minor* of a square matrix is the determinant of a (leading) principal submatrix.

Principal minors become relevant when one wants to test whether a given symmetric matrix is positive semidefinite (for definitions, see Section 4.2).

THEOREM 9 A symmetric matrix is positive semidefinite if and only if all its principal minors are non-negative.

Notice that non-negativity of the *leading* principal minors is necessary but not sufficient for positive semidefiniteness. For example, the 3×3 diagonal matrix **D** with diagonal entries $1, 0, -1$ is not positive semidefinite but its leading principal minors are 1, 0, 0. We emphasize this because Theorem 9 is often confused with the next theorem, which is about positive *definite* matrices.

THEOREM 10 For a symmetric matrix **S**, the following three statements are equivalent.

(α) **S** is positive definite.

(β) All principal minors of **S** are positive.

(γ) All leading principal minors of **S** are positive.

A symmetric matrix **S** is said to be negative definite (semidefinite) if $-\mathbf{S}$ is positive definite (semidefinite).

This brings us at last to the concavity test. A standard theorem in calculus states that the Hessian matrix of a C^2 function is symmetric. (For a rigorous discussion see Bartle, 1976, pp. 367–9; the result is sometimes called Young's Theorem, but Bartle attributes it to H. A. Schwarz of inequality fame.) Thus Theorem 10.2.6 states that a C^2 function f on an open convex set U in R^n is concave if and only if its Hessian matrix $\mathbf{H}(\mathbf{x})$ is negative semidefinite at every point $\mathbf{x}$ of U. By Theorem 9, f is concave if and only if all principal minors of $-\mathbf{H}(\mathbf{x})$ are non-negative functions on U.

Example 3 We define the function

$$f(x_1, x_2) = \sqrt{x_1} \cdot \ln(1 + x_2)$$

on the positive orthant in R^2 and show that it is not concave.

Let $\mathbf{S}(\mathbf{x}) = -\mathbf{H}(\mathbf{x})$, where $\mathbf{H}$ is the Hessian. The principal minors of $\mathbf{S}(\mathbf{x})$ are

$$s_{11}(\mathbf{x}) = \tfrac{1}{4} x_1^{-3/2} \ln(1 + x_2)$$
$$s_{11}(\mathbf{x}) = x_1^{1/2}(1 + x_2)^{-2}$$

and

$$\det \mathbf{S}(\mathbf{x}) = (\ln(1 + x_2) - 1)/(4x_1(1 + x_2)^2).$$

Now s_{11} and s_{22} are positive but det $\mathbf{S}$ is negative when $x_2 < e - 1$.

Appendix D: Input-Output Analysis

The main interest of input–output analysis lies in its applications. The rather small amount of theory involved uses two of the main concepts of this book: Gaussian elimination and linear programming. So it is worthwhile to devote a few pages to it.

Consider a closed economy (no foreign trade) with n producible goods. There may also be non-produced primary factors.

Assume *no joint production*: thus industry j $(= 1, \ldots, n)$ produces positive quantities only of good j, using quantities of the other goods and primary factors as inputs. Also assume *constant returns to scale* and *fixed coefficients in production.* Let a_{ij} denote the quantity of good i used as input in industry j per unit of j produced.

Denote the *gross output* of good i, including that fed back into the system as intermediate input, by x_i. Denote the *net output* of good i, available for consumption or accumulation, by y_i. Then

y_i = gross output of i
less quantity of i required as input in industry 1
less quantity of i required as input in industry 2
less ...
less quantity of i required as input in industry n

$$= x_i - a_{i1}x_1 - a_{i2}x_2 - \cdots - a_{in}x_n$$

In matrix terminology,

$$\mathbf{y} = (\mathbf{I} - \mathbf{A})\mathbf{x} \qquad (1)$$

where $\mathbf{y}$ is the net output vector, $\mathbf{x}$ is the gross output vector and

$\mathbf{A} = (a_{ij})$ is the *input–output matrix*. In practice, one might want to find how much gross output of each good is required to yield a given bundle of net outputs. This means solving (1) for $\mathbf{x}$ given a non-negative $\mathbf{y}$, and hoping for a non-negative solution.

As in Section 5.3, we define an n-vector $\mathbf{z}$ to be non-negative ($\mathbf{z} \geq \mathbf{0}$) if all its components are non-negative. As in Section 10.1, we define an n-vector $\mathbf{z}$ to be positive ($\mathbf{z} > \mathbf{0}$) if all its components are positive.

A *Leontief matrix* is a square matrix whose off-diagonal entries are all non-positive. The matrix $\mathbf{I} - \mathbf{A}$ in (1) is a Leontief matrix and has the additional property that its diagonal entries do not exceed 1. The latter property is *not* part of the definition of a Leontief matrix. We insist on this because we want 'Leontief' to be a property that is preserved when columns are multiplied by arbitrary positive scalars: why we want to do this will become clear later.

The main theorem about Leontief matrices runs as follows.

THEOREM 1 If $\mathbf{B}$ is an $n \times n$ Leontief matrix, the following three statements are equivalent.

(α) For *some* $\mathbf{y}^0 > \mathbf{0}$ the system $\mathbf{Bx} = \mathbf{y}^0$ has a positive solution $\mathbf{x}$.
(β) For *every* $\mathbf{y} \geq \mathbf{0}$ the system $\mathbf{Bx} = \mathbf{y}$ has a non-negative solution $\mathbf{x}$.
(γ) $\mathbf{B}$ is nonsingular and all entries of $\mathbf{B}^{-1}$ are non-negative.

PROOF If (β) holds there exist non-negative vectors $\mathbf{v}_1, \ldots, \mathbf{v}_n$ such that $\mathbf{Bv}_j$ is column j of $\mathbf{I}_n$: this implies (γ). If (γ) holds then $\mathbf{B}^{-1}$ has all entries non-negative and no row consisting entirely of zeroes. Hence $\mathbf{B}^{-1}\mathbf{y}^0 > \mathbf{0}$ for all positive $\mathbf{y}^0$ and (α) is true. It remains to show (α) implies (β).

Let $\mathbf{y}^0$ be as in (α) and let $\mathbf{y} \geq \mathbf{0}$. We attempt to solve the system $\mathbf{Bx} = \mathbf{y}$ by Gaussian elimination. Since (α) holds, and $y_1^0 > 0$ and $b_{1j} \leq 0$ for $j > 1$, $b_{11} > 0$. Hence we can perform one elimination step without row interchange. This yields the array

$$\left.\begin{array}{cccc} b_{11} & b_{12} & \cdots & b_{1n} \\ 0 & c_{22} & \cdots & c_{2n} \\ & & \vdots & \\ 0 & c_{n2} & \cdots & c_{nn} \end{array}\right| \begin{array}{c} y_1 \\ z_2 \\ \vdots \\ z_n \end{array}$$

Since $b_{11} > 0$ and $b_{1j} \leq 0$ for $j > 1$ and $b_{i1} \leq 0$ for $i > 1$, we have $z_i \geq y_i$ and $c_{ij} \leq b_{ij}$ for all $i > 1$ and all $j > 1$. Thus (c_{ij}) is a Leontief matrix and

$$(y_1 \colon \mathbf{z}) \geq \mathbf{y} \geq \mathbf{0}.$$

In particular, $z_2^0 \geq y_2^0 > 0$. But then $c_{22} > 0$ by (α) and we can perform another elimination step without row interchange.

Proceeding in this way we see that the system $\mathbf{Bx} = \mathbf{y}$ has an equivalent system $\mathbf{Ux} = \mathbf{w}$, where $\mathbf{w} \geq \mathbf{0}$ and $\mathbf{U}$ is an upper triangular Leontief matrix with positive diagonal entries. By back-substitution, $\mathbf{x} \geq \mathbf{0}$. QED

A Leontief matrix **B** which satisfies one and hence all of the properties (α), (β), (γ) of Theorem 1 is said to be *workable*. The proof of Theorem 1 gives a simple test for workability. Apply Gaussian elimination to **B**: if the rth diagonal entry after $r - 1$ elimination steps is positive for all $r = 1, \ldots, n$, then **B** is workable; if not, not. This test can be given a determinantal interpretation, in which case it is called the *Hawkins–Simon condition*: a Leontief matrix is workable if and only if all its leading principal minors are positive. But the test itself is pure Gaussian elimination, and no knowledge of determinants is required for its understanding or its practice.

Returning to the economic model, we say that the input–output matrix **A** is *productive* if the Leontief matrix $\mathbf{I} - \mathbf{A}$ is workable. Conditions (α) and (β) of Theorem 1 give two equivalent ways of describing a productive input–output matrix: **A** is productive if *some* positive net output vector is producible, given sufficient gross outputs; and **A** is productive if *any* non-negative net output vector is producible, given sufficient gross outputs. The equivalence of these two conditions is really rather remarkable: clearly it depends heavily on the no-joint-production assumption.

We now turn to the special case where there is only one primary factor, which we call *labour*. Suppose that for each industry j, some *positive* quantity c_j of labour is required per unit of *gross* output of j. Let **A** be the input–output matrix (assumed productive), **B** the workable Leontief matrix $\mathbf{I} - \mathbf{A}$, and let the square matrix **G** be defined by

$$g_{ij} = b_{ij}/c_j.$$

Thus g_{ij} is the net output of good i in industry j per unit of labour employed in industry j.

We call **G** the *normalised technology matrix*: it is a workable Leontief matrix. The Leontief property is obvious. To prove workability notice that $\mathbf{G} = \mathbf{BD}^{-1}$ where **D** is the diagonal matrix

$$\operatorname{diag}(c_1, \ldots, c_n).$$

Then $\mathbf{G}^{-1} = \mathbf{DB}^{-1}$, all of whose entries are non-negative.

Suppose that for $j = 1, \ldots, n$, z_j units of labour are allocated to industry j: then gross output of j is z_j/c_j. Thus net outputs $y_1, \ldots, y_n$ are given by

$$y_i = \sum_{j=1}^{n} b_{ij}(z_j/c_j) = \sum_{j=1}^{n} g_{ij} z_j$$

so $\mathbf{y} = \mathbf{Gz}$.

Let L denote the total quantity of labour available and let $\mathbf{e}$ be the n-vector consisting entirely of 1's. Then the net output vector $\mathbf{y}$ is on the economy's production frontier if and only if there exists a non-negative labour allocation vector $\mathbf{z}$ such that $\mathbf{y} = \mathbf{Gz}$ and $\mathbf{e}^{\mathrm{T}}\mathbf{z} = L$.

Since all entries of $\mathbf{G}^{-1}$ are non-negative, so are all entries of

$$(\mathbf{G}^{\mathrm{T}})^{-1} = (\mathbf{G}^{-1})^{\mathrm{T}}.$$

Since this matrix is nonsingular, we can define a *positive* n-vector $\mathbf{p}$ by

$$\mathbf{p} = (\mathbf{G}^{\mathrm{T}})^{-1}\mathbf{e}. \qquad (2)$$

Then we may give the following expression for the production frontier T:

$$T = \{\mathbf{y} \in R^n \colon \mathbf{y} \geq \mathbf{0} \quad \text{and} \quad \mathbf{p}^{\mathrm{T}}\mathbf{y} = L\}. \qquad (3)$$

As we might expect, $\mathbf{p}$ has a price interpretation. To see this, suppose that the prices of the goods are $q_1, \ldots, q_n$ and that the price of labour is w, all in terms of some common numéraire. Then net profit in industry j per unit of gross output of j is

$$q_j - \left(\sum_{i=1}^{n} q_i a_{ij}\right) - wc_j = \left(\sum_{i=1}^{n} q_i b_{ij}\right) - wc_j$$

$$= \left[\left(\sum_{i=1}^{n} q_i g_{ij}\right) - w\right] c_j.$$

The expression in square brackets is the jth component of $\mathbf{G}^T\mathbf{q} - w\mathbf{e}$. We know from (2) that this vector is zero if and only if $w^{-1}\mathbf{q} = \mathbf{p}$. Thus $\mathbf{p}$ is that vector of prices, with labour as numéraire, which makes all industries just break even.

Summarising these results, we see that the production frontier T, given by (3), is flat: it is the intersection of a hyperplane with the non-negative orthant. The competitive price vector $\mathbf{p}$ is determined entirely by the technology, and depends not at all on the structure of consumer demand.

These results depend on the assumptions of constant returns to scale, no joint production, and the existence of only one primary factor (labour). Interestingly, they do not require the assumption of fixed coefficients that we have been using so far. The rest of this appendix explains why this is so.

We continue to assume that there are n producible goods, constant returns to scale, and only one primary factor (labour). We introduce choice of technique as follows. Suppose that there are $r(> n)$ possible production processes, each requiring the use of labour. We define the $n \times r$ matrix $\mathbf{H}$ by letting h_{ik} be the net output of good i in process k per unit of labour employed on k.

We impose again the assumption of no joint production by requiring that for each process k there is exactly one integer j, depending on k, such that

$$h_{jk} > 0 \geq h_{ik} \quad \text{for all } i \neq j.$$

Process k is then said to be a process for industry j.

A *Leontief selection* is a list of processes $(k_1, \ldots, k_n)$, one for each industry. Given such a selection we may choose a normalised technology matrix $\mathbf{G}$ such that column j of $\mathbf{G}$ is column k_j of $\mathbf{H}$ for $j = 1, \ldots, n$. The Leontief selection is said to be workable if $\mathbf{G}$ is workable.

From now on we make the productivity assumption that there is at least one workable Leontief selection.

Example 1 Let there be 3 goods and 6 processes, with

$$\mathbf{H} = \begin{pmatrix} 3 & 2 & -1 & 0 & -1 & -8 \\ -1 & 0 & 2 & 3 & -1 & -4 \\ -1 & -1 & 0 & -1 & 1 & 5 \end{pmatrix}.$$

Then processes 1 and 2 are for industry 1, processes 3 and 4 for industry 2 and processes 5 and 6 for industry 3. So three Leontief selections are

$$(1, 3, 5), \quad (2, 4, 5), \quad (1, 4, 6).$$

The normalised technology matrix for the third selection is

$$\begin{pmatrix} 3 & 0 & -8 \\ -1 & 3 & -4 \\ -1 & -1 & 5 \end{pmatrix}.$$

The reader may easily verify by Gaussian elimination that this matrix is workable.

Returning to the general analysis, suppose that total labour supply is L. If z_k units of labour are allocated to process k for each k, the net output vector is $\mathbf{Hz}$. For such an allocation to be feasible we need $\mathbf{f}^{\mathrm{T}}\mathbf{z} \leq L$, where $\mathbf{f}$ is the r-vector of 1's: $\mathbf{e}$ continues to denote the n-vector of 1's. We say that the n-vector $\mathbf{y}$ is a *producible* net output vector if $\mathbf{y} \geq \mathbf{0}$ and there exists an r-vector $\mathbf{z}$ such that

$$\mathbf{y} = \mathbf{Hz}, \quad \mathbf{f}^{\mathrm{T}}\mathbf{z} \leq L \quad \text{and} \quad \mathbf{z} \geq \mathbf{0}. \tag{4}$$

The set of all producible $\mathbf{y}$ is called the *producible set* and denoted by Y. The member $\mathbf{y}$ of Y is said to be *efficient* if there is no $\mathbf{y}^*$ in Y, other than $\mathbf{y}$ itself, for which $\mathbf{y}^* \geq \mathbf{y}$. The set of efficient net output vectors is the economy's production frontier, which we again denote by T. It follows from the assumption of constant returns to scale that if $\mathbf{y}$ is efficient in this sense, and $\mathbf{z}$ satisfies (4) for this $\mathbf{y}$, then $\mathbf{f}^{\mathrm{T}}\mathbf{z} = L$ (full employment).

We say that the net output vector $\mathbf{y}$ is producible via the Leontief selection S if there exists $\mathbf{z}$ satisfying (4) such that $z_k = 0$ for all k not in S.

THEOREM 2 (Non-Substitution Theorem)
Under the assumptions made above, there is exactly one n-vector $\mathbf{p}$ such that:
(I) the production frontier T is the set

$$\{\mathbf{y} \in R^n: \mathbf{y} \geq \mathbf{0} \quad \text{and} \quad \mathbf{p}^{\mathrm{T}}\mathbf{y} = L\}.$$

Also $\mathbf{p} > \mathbf{0}$ and there exists a Leontief selection S such that:
(II) any member of T is producible via S using L units of labour;
(III) with goods prices $\mathbf{p}$ in terms of labour as numéraire, processes in S break even and no process makes a positive profit.

PROOF Let $\mathbf{b}$ be any positive n-vector. We set up a linear programme to minimise the quantity of labour required to produce the net output vector $\mathbf{b}$. This is

$$\text{(M)} \quad \text{minimise } \mathbf{f}^{\mathrm{T}}\mathbf{z} \quad \text{subject to } \mathbf{Hz} = \mathbf{b} \text{ and } \mathbf{z} \geq \mathbf{0}.$$

Since at least one workable Leontief selection exists, (M) is feasible. Since $\mathbf{f} > \mathbf{0}$, (M) is bounded. So by Theorem 8.2.2, there exists a solution vector $\hat{\mathbf{z}}$ for (M) with at most n positive components. By the no-joint-production assumption, $\hat{\mathbf{z}}$ has exactly n positive components.

We may therefore choose a workable Leontief selection S and a positive n-vector $\hat{\mathbf{x}}$ such that

$$\mathbf{G}\hat{\mathbf{x}} = \mathbf{H}\hat{\mathbf{z}} = \mathbf{b}$$

where $\mathbf{G}$ is the normalised technology matrix for S.

The dual of (M) is

$$\text{(M*)} \quad \text{maximise } \mathbf{b}^{\mathrm{T}}\mathbf{q} \quad \text{subject to } \mathbf{H}^{\mathrm{T}}\mathbf{q} \leq \mathbf{f}.$$

Let $\mathbf{p}$ be a solution vector for (M*). By the Complementarity Theorem,

$$\mathbf{H}^{\mathrm{T}}\mathbf{p} \leq \mathbf{f} \quad \text{and} \quad \mathbf{G}^{\mathrm{T}}\mathbf{p} = \mathbf{e}. \tag{5}$$

Since $\mathbf{G}$ is workable, $\mathbf{p} = (\mathbf{G}^{\mathrm{T}})^{-1}\mathbf{e}$. Hence $\mathbf{p} > \mathbf{0}$.

Clearly, there can be at most one $\mathbf{p}$ satisfying part (I) of the theorem. We now show that the $\mathbf{p}$ of (5) does the trick.

Let $\mathbf{y}$ be any non-negative n-vector and let $\mathbf{x} = \mathbf{G}^{-1}\mathbf{y}$, $\lambda = \mathbf{p}^{\mathrm{T}}\mathbf{y}$. Then $\lambda = \mathbf{e}^{\mathrm{T}}\mathbf{x}$. We consider three cases.

(i) Let $\lambda > L$. If $\mathbf{y} = \mathbf{Hz}$ and $\mathbf{z} \geq \mathbf{0}$ then

$$\mathbf{f}^{\mathrm{T}}\mathbf{z} \geq \mathbf{p}^{\mathrm{T}}\mathbf{Hz} = \lambda > L$$

so $\mathbf{y}$ is not producible.

(ii) Let $\lambda = L$. Then $\mathbf{y} = \mathbf{Gx}$ and $\mathbf{e}^{\mathrm{T}}\mathbf{x} = L$ so $\mathbf{y} \in Y$. If $\mathbf{y}^* \geq \mathbf{y}$ and $\mathbf{y}^* \neq \mathbf{y}$ then $\mathbf{p}^{\mathrm{T}}\mathbf{y}^* > L$ by positivity of $\mathbf{p}$ so $\mathbf{y}^*$ is not producible, by case (i). Thus $\mathbf{y} \in T$.

(iii) Let $\lambda < L$. Then for sufficiently small positive α

$$\mathbf{e}^{\mathrm{T}}\mathbf{x} + \alpha\mathbf{e}^{\mathrm{T}}\mathbf{G}^{-1}\mathbf{e} < L$$

so $\mathbf{y} + \alpha\mathbf{e} \in Y$. Thus $\mathbf{y}$ is not an efficient point of Y.

The statement (I) follows from cases (i), (ii), (iii). We have already shown that $\mathbf{p} > \mathbf{0}$. The analysis of case (ii) establishes (II). (III) follows from (5). QED

The theorem gets its name from statement (II): the economy can get by (efficiently) on just one Leontief selection, regardless of consumers' preferences.

The proof of the theorem yields an algorithm for finding the competitive price vector $\mathbf{p}$ and the efficient Leontief selection S. Recall that $\mathbf{p}$ is the unique solution vector for (M*) and $\mathbf{p} > \mathbf{0}$. Thus $\mathbf{p}$ is the unique solution vector for the linear programme

$$\text{(N)} \quad \text{maximise } \mathbf{b}^{\mathrm{T}}\mathbf{q} \quad \text{subject to } \mathbf{H}^{\mathrm{T}}\mathbf{q} \leq \mathbf{f} \quad \text{and} \quad \mathbf{q} \geq \mathbf{0},$$

so to find $\mathbf{p}$ we solve (N) by the simplex method. To find an efficient Leontief selection we choose processes, one for each industry, which break even at prices $\mathbf{p}$. This is done by choosing those k for which $n + k$ is a column label in the final scheme.

The economic interpretation of (N) is that we choose prices to maximise the value (in terms of labour) of producing $\mathbf{b}$, subject to the constraint that no process makes a positive profit. Recall from the proof that $\mathbf{b}$ can be any positive n-vector: a natural choice for the algorithm is $\mathbf{b} = \mathbf{e}$.

Example 2 Let $\mathbf{H}$ be as in Example 1. Solving (N), with $\mathbf{b} = \mathbf{e}$, by the simplex method we obtain the scheme

		4	8	6
	11	3	4	4
1	3			
5	1			
3	6			
2	2			
7	1			
9	3			

Hence $p_1 = 3$, $p_2 = 2$ and $p_3 = 6$. The production frontier is

$$\{\mathbf{y} \in R^3 \colon \mathbf{y} \geq \mathbf{0} \quad \text{and} \quad 3y_1 + 2y_2 + 6y_3 = L\},$$

where L is total labour supply.

The column labels are 4, 8, 6, so processes 1, 5, 3 break even at $\mathbf{p}$. The efficient Leontief selection is process 1 for industry 1, 3 for 2, and 5 for 3.

The discussion of choice of technique leading to Theorem 2 was in terms of a choice from a finite menu, the columns of $\mathbf{H}$. The extension to an infinite menu (for example, 'neoclassical' production functions) is easy. The assumptions we need are

(a) constant returns to scale,
(b) no joint production,
(c) one primary factor, required for every process,
(d) existence of a positive efficient net output vector,
(e) convexity.

What convexity means in this context is that labour can be allocated across processes in any way we like.

To prove Theorem 2 under these assumptions, let $\hat{\mathbf{y}}$ be a positive efficient net output vector; let S be a Leontief selection, with associated normalised technology matrix $\mathbf{G}$, which can produce $\hat{\mathbf{y}}$ using L units of labour; and let $\mathbf{p} = (\mathbf{G}^{\mathrm{T}})^{-1}\mathbf{e}$. Let $\mathbf{h}$ be any process for any industry, and apply the first part of the proof of Theorem 2, with $\mathbf{b} = \hat{\mathbf{y}}$, to the truncated economy with process matrix $\mathbf{H} = (\mathbf{G}, \mathbf{h})$. We infer that $\mathbf{p}^{\mathrm{T}}\mathbf{h} \leq 1$. Since this is so for any $\mathbf{h}$, the rest of the proof goes through as before.

Notes on Further Reading

Mathematics

Two good textbooks on linear algebra are Halmos (1958) and Strang (1976). They are as different as their titles suggest, with Strang emphasising practical matters. Chapters 5–7 of Strang are particularly good on eigenvalues, the main topic in linear algebra omitted from this book.

Gass (1975) discusses linear programming in detail. Luenberger (1973) contains useful material on linear programming and a great deal on computational aspects of nonlinear programming.

The last two chapters of this book touched on the field of convex analysis: the standard work on this is Rockafellar (1970).

Econometrics

The algebra of Chapter 4 of this book is applied in all the main textbooks on econometrics. Three of these, in ascending order of difficulty, are Wallis (1972), Maddala (1977) and Malinvaud (1970).

Economics

The excellent little book by Dixit (1976) discusses some of the same topics as this book, with some economic applications. Dixit's approach is very different from mine in that he emphasises a general understanding of concepts, whereas I have tried to impart a working knowledge of the relevant mathematics. The last three chapters of Dixit are a useful introduction to optimisation over time.

A major field of application of mathematical programming is development planning. For an introductory survey see Manne (1974).

Many books on linear programming contain a section on two-person zero-sum matrix games. The discovery by Princeton mathematicians in the late 1940s that the main theorem on such games is a theorem about duality in linear programming was an important contribution both to game theory and to mathematical programming. But the two-person zero-sum case is very special, and it is the generalisation to Nash equilibria of non-cooperative games that seems to be of more economic interest: see Chapters 1 and 7 of Friedman (1977).

The method of complementary solutions has been extended far beyond linear programming. One important extension to economics consists of algorithms for computing equilibrium prices: see Scarf (1973) for the theory and Miller and Spencer (1977) for a practical application. Lemke and Howson (1964) discuss complementary solutions in game theory: for generalisations see the papers by Cottle and Dantzig, Eaves, and Shapley and Scarf in Dantzig and Eaves (1974). Better late than never, we must mention the fact that G. B. Dantzig is one of the two fathers of linear programming; the other, working quite independently in the late 1930s, is the Soviet mathematician and economist, L. V. Kantorovich.

An approach similar to the transportation problem is applied by Rosen (1978) to some classical problems in labour economics.

The Kuhn–Tucker Theorem and related issues loom large in the theory of economic planning (Heal, 1973) and in modern work on public finance. The latter field is surveyed by Atkinson and Stiglitz (1980); for contributions emphasising the mathematics, see Mirrlees (1976) and Diewert (1978). Pissarides (1980) gives a nice application of the Kuhn–Tucker Theorem to the analysis of labour contracts.

For an introduction to input–output analysis by its creator, see Leontief (1968); see also Leontief and Ford (1972) for an application to environmental pollution. Arrow and Starrett (1973) discuss the place of input–output analysis within the general context of price theory.

References

K. J. Arrow and D. A. Starrett, 'Cost and Demand—Theoretical Approaches to the Theory of Price Determination', in J. R. Hicks and W. Weber (eds), *Carl Menger and the Austrian School of Economics* (Oxford University Press, 1973).

A. B. Atkinson and J. E. Stiglitz, *Lectures on Public Economics* (McGraw-Hill, 1980).

R. G. Bartle, *The Elements of Real Analysis*, 2nd ed. (Wiley, 1976).

R. G. Bland, 'New Finite Pivoting Rules for the Simplex Method', *Mathematics of Operations Research*, II (1977) 103–7.

D. Cass, 'Duality: a Symmetric Approach from the Economist's Vantage Point', *Journal of Economic Theory*, VII (1974) 272–95.

A. C. Chiang, *Fundamental Methods of Mathematical Economics*, 2nd ed. (McGraw-Hill, 1974).

G. B. Dantzig and B. C. Eaves (eds), *Studies in Optimization* (Mathematical Association of America, 1974).

W. Diewert, 'Optimal Tax Perturbations', *Journal of Public Economics*, X (1978) 1–24.

A. K. Dixit, *Optimisation in Economic Theory* (Oxford University Press, 1976).

J. W. Friedman, *Oligopoly and the Theory of Games* (North-Holland, 1977).

S. I. Gass, *Linear Programming*, 4th ed. (McGraw-Hill, 1975).

P. R. Halmos, *Finite Dimensional Vector Spaces* (Van Nostrand, 1958).

G. M. Heal, *The Theory of Economic Planning* (North-Holland, 1973).

H. W. Kuhn and A. W. Tucker, 'Nonlinear Programming', in J. Neyman (ed.), *Proceedings of the Second Berkeley Symposium on Mathematical Statistics and Probability* (University of California Press, 1951).

C. E. Lemke and J. T. Howson, Jr, 'Equilibrium Points of Bimatrix Games', *Journal of the Society of Industrial and Applied Mathematics*, XII (1964) 413–23.

W. Leontief, 'Input–Output Analysis' in D. Sills (ed.), *International Encyclopaedia of the Social Sciences*, vol. VII (Macmillan and The Free Press, 1968), pp. 344–54.

W. Leontief and Daniel Ford, 'Air Pollution and the Economic Structure: Empirical Results of Input–Output Computations', in A. Brody and A. P. Carter (eds), *Input–Output Techniques* (North-Holland, 1972).

D. G. Luenberger, *Introduction to Linear and Nonlinear Programming* (Addison-Wesley, 1973).

E. J. McShane, 'The Lagrange Multiplier Rule', *American Mathematical Monthly*, LXXX (1973) 922–5.

G. S. Maddala, *Econometrics* (McGraw-Hill, 1977).

E. Malinvaud, *Statistical Methods of Econometrics*, 2nd ed. (North-Holland, 1970).

A. S. Manne, 'Multi-Sector Models for Development Planning: a Survey', in M. D. Intriligator and D. A. Kendrick (eds), *Frontiers of Quantitative Economics*, vol. II (North-Holland, 1974).

M. H. Miller and J. E. Spencer, 'The Static Economic Effects of the U.K. Joining the E.E.C.: A General Equilibrium Approach', *Review of Economic Studies*, XLIV (1977) 71–93.

J. A. Mirrlees, 'Optimal Tax Theory: a Synthesis', *Journal of Public Economics*, VI (1976) 327–58.

C. A. Pissarides, 'Contract Theory, Temporary Layoffs and Unemployment: a Critical Assessment', in D. A. Currie and W. Peters (eds), *Contemporary Economic Analysis*, vol. III (Croom Helm, 1980).

R. T. Rockafellar, *Convex Analysis* (Oxford University Press, 1970).

S. Rosen, 'Substitution and Division of Labour' *Economica*, XLV (1978) 235–50.

H. Scarf, *The Computation of Economic Equilibria* (Yale University Press, 1973).

G. Strang, *Linear Algebra and its Applications* (Academic Press, 1976).

A. W. Tucker, 'Combinatorial Algebra of Linear Programs' in Dantzig and Eaves (1974).

K. F. Wallis, *Introductory Econometrics* (Gray-Mills, 1972).

Index